Handbook of
Geriatrics

Associate Editors:

Dalton M. Benson, M.D.
Assistant Professor
Division of Gerontology and Geriatric Medicine
Department of Medicine
New York Medical College
Valhalla, New York
and Chief of Intermediate Medicine and Geriatrics
Franklin Delano Roosevelt Veterans Administration Hospital
Montrose, New York

Krishan L. Gupta, M.D.
Associate Professor
Division of Gerontology and Geriatric Medicine
Department of Medicine
New York Medical College
Valhalla, New York
and Associate Chief of Medical Services
Ruth Taylor Geriatric and Rehabilitation Institute
Westchester County Medical Center
Hawthorne, New York

Handbook of
Geriatrics

Edited by

Steven R. Gambert, M.D., F.A.C.P.

Professor of Medicine
Director, Division of Gerontology and Geriatric Medicine
Department of Medicine
and Director, Center for Aging and Adult Development
New York Medical College
Valhalla, New York

Director of Geriatrics
Westchester County Medical Center
and Chief of Medical Services
Ruth Taylor Geriatric and Rehabilitation Institute
Valhalla, New York

Plenum Medical Book Company
New York and London

Library of Congress Cataloging in Publication Data

Handbook of geriatrics.

Includes bibliographies and index.
1. Geriatrics—Handbooks, manuals, etc. I. Gambert, Steven R., 1949-   . II. Ti
tle. [DNLM: 1. Geriatrics—handbooks. WT 39 H2365]
RC952.55.H39   1986   618.97   86-2535:
ISBN 0-306-42336-7

First Printing—March 1987
Second Printing—November 1989

© 1987 Plenum Publishing Corporation
233 Spring Street, New York, N.Y. 10013

Plenum Medical Book Company is an imprint of Plenum Publishing Corporation

Printed in the United States of America

Contributors

Dalton M. Benson, M.D., Assistant Professor, Division of
Gerontology and Geriatric Medicine, Department of Medicine,
New York Medical College, Valhalla, New York 10595; and
Chief of Intermediate Medicine and Geriatrics, Franklin
Delano Roosevelt Veterans Administration Hospital, Montrose,
New York 10548

Daniel J. Cameron, M.D., M.P.H., Assistant Professor, Division
of Gerontology and Geriatric Medicine, Department of
Medicine, New York Medical College, Valhalla, New York
10595; and Director, Dementia Unit, Ruth Taylor Geriatric and
Rehabilitation Institute, Westchester County Medical Center,
Hawthorne, New York 10532

Brad Dworkin, M.D., Associate Professor, Sarah C. Upham
Division of Gastroenterology, Department of Medicine, New
York Medical College, Valhalla, New York 10595; and
Director, Nutrition Support Service, Westchester County
Medical Center, Hawthorne, New York 10532

Jeffrey Escher, M.D., Assistant Professor and Director of
Geriatric Clinics, Division of Gerontology and Geriatric
Medicine, Department of Medicine, New York Medical
College, Valhalla, New York 10595; and Staff Geriatrician,
Ruth Taylor Geriatric and Rehabilitation Institute, Westchester
County Medical Center, Hawthorne, New York 10532

Steven R. Gambert, M.D., Professor of Medicine, Director,
Division of Gerontology and Geriatric Medicine, Department
of Medicine, and Director, Center for Aging and Adult

Development, New York Medical College, Valhalla, New York 10595; and Director of Geriatrics, Westchester County Medical Center, and Chief of Medical Services, Ruth Taylor Geriatric and Rehabilitation Institute, Hawthorne, New York 10532

Krishan L. Gupta, M.D., Associate Professor, Division of Gerontology and Geriatric Medicine, Department of Medicine, New York Medical College, Valhalla, New York 10595; and Associate Chief of Medical Services, Ruth Taylor Geriatric and Rehabilitation Institute, Westchester County Medical Center, Hawthorne, New York 10532

Crossley O'Dell, M.D., Fellow in Geriatric Medicine, Westchester County Medical Center, and Clinical Instructor, Division of Gerontology and Geriatric Medicine, Department of Medicine, New York Medical College, Valhalla, New York 10595

Lea Scroggs, M.D., Clinical Instructor, Division of Gerontology and Geriatric Medicine, Department of Medicine, New York Medical College, Valhalla, New York 10595

Katherine Small, M.D., Assistant Professor, Division of Infectious Diseases, Department of Medicine, New York Medical College, Valhalla, New York 10595

Panayiotis D. Tsitouras, M.D., Associate Professor, Division of Gerontology and Geriatric Medicine, Department of Medicine, New York Medical College, Valhalla, New York 10595

Mark Zimmerman, M.D., Assistant Professor, Division of Gerontology and Geriatric Medicine, Department of Medicine, New York Medical College, Valhalla, New York 10595; and Staff Geriatrician, Ruth Taylor Geriatric and Rehabilitation Institute, Westchester County Medical Center, Hawthorne, New York 10532

Preface

As our population continues to age, health professionals are being
called on to care for an ever-increasing number of elderly patients.
A thorough understanding of what constitutes normal aging versus
age-prevalent illness is essential. In addition, the atypical and
nonspecific presentation of illness commonly encountered when
caring for an older patient must be expected and watched for
carefully.

In recent years, the health professional has been exposed to an
exponentially increasing number of publications attempting to teach
geriatric principles. To date, few publications lend themselves to
use by the busy practitioner, student, or nurse in search of
immediate facts, flow sheets, and clinically applicable data.

It was felt that the health professional would benefit greatly
from a book based on the concept of a ready-reference "handbook," with chapters filled with tables, flow sheets, and listings
similar in scope to those in a well-presented lecture series. Our goal
was to create a geriatrics handbook that would have value at the
bedside as well as in the classroom. It is to this end that the
contributors dedicated their efforts.

The end product, we believe, is appealing to the busy physician caring for the elderly patient. In addition, the physician-in-
training, the physician assistant, the nurse practitioner, and the
clinically astute nurse will find this handbook to be a ready
reference that will be well worth the pocket space it takes up. The
teacher may also find the book's format of use in preparing lectures
on a wide variety of problems relating to the health care of the
elderly. In fact, many of the tables and flow sheets used in this
handbook were originally used in lectures personally given throughout the country on a wide range of geriatric topics.

Appreciation is given to our patients, families, mentors, and
readers, without whom this book would not have been possible.

Special appreciation is given to the editorial staff at Plenum
Publishing Corporation for their continued support in bringing
knowledge to the forefront of clinical practice, and to Arlene
Moore, my administrative assistant, and Doreen Derry for providing
assistance in preparing the manuscript. Most of all, appreciation is
given to my family, Gry, Iselin, and Christopher, for giving of
their own time and spirit to allow me to complete this book in a
timely fashion.

Steven R. Gambert

Valhalla, New York

Contents

Chapter 6
Genitourinary Problems in the Elderly

Chapter 7
Endocrine and Metabolic Problems in the Elderly

Chapter 8
Dermatologic Problems in the Elderly

Chapter 12
Infectious Diseases in the Elderly

Chapter 13
Selected Neurological Problems in the Elderly

Chapter 14
Preventive Health Care for the Elderly

Chapter 15
Patient Education Handouts

1

Aging: An Overview

I. THE ELDERLY: A DEMOGRAPHIC IMPERATIVE

At present there are approximately 28 million persons over age 65 in the United States, constituting 11.5% of the total population. It has been estimated that by the year 2020, approximately 20% of America's population will be over the age of 65. Perhaps more important, however, is the shift toward an "older" geriatric population; by the turn of the century, there will be more persons over the age of 75 than those aged 65–75. The elderly, especially those over the age of 75, constitute the single largest group of consumers of health services in the country and expend 25% of all health care dollars. In fact, one-third of all acute hospital beds and 90% of long-term care beds are occupied by persons over 65 years of age. The elderly use over 25% of prescription and an even higher percentage of over-the-counter medications.

U.S. Population by Age (1980)		
Age (years)	Population	Percentage of total
Under 5	16,344,407	7.2
5–15	34,938,053	15.4
15–44	105,181,100	46.4
45–64	44,497,132	19.6
≥65	25,544,133	11.3
Total	226,504,025	100.0

Percentage of Population in Elderly Age Groups—Past and Projected

Year	>65 Years (%)	>75 Years (%)
1900	4.0	1.2
1910	4.3	1.3
1940	6.8	2.0
1950	8.1	2.6
1960	9.2	3.1
1970	9.9	3.6
1980	11.3	4.4
1990	12.7	4.9
2000	13.1	5.2
2010	13.9	5.5
2030	21.1	7.7

Life expectancy has increased significantly throughout this century from that of 48 and 52 years, respectively, for a boy and girl born in 1900. Today a male child born in this country can expect to live to the age of 72; a newborn female has a life expectancy of approximately 79 years.

It is important to view the older person, however, as a "survivor," one who did not die from childhood illness or age-prevalent disease. At age 70, a woman has a projected life span of an additional 17 years, and a man an additional 12 years. Despite advances in prolonging the expected life span (an age that at least 50% of the population can "expect" to live to), little change has occurred in maximum life span (maximum longevity for the species), currently set at 100–120 years. There are currently 40,000 centenarians in the United States. Socioeconomic advancement is the single most important factor in the increase in "expected" life span. Other factors include improved nutrition, immunization programs, health maintenance, and sanitation. In addition, there has been major advancement in curing infectious diseases, responsible for millions of deaths in the early part of this century. Acute diseases have consistently been replaced by chronic disorders. At present, the two leading causes of death are cardiovascular disease and cancer.

The three periods of the life span include growth and development, maturation, and senescence. There are, however, no clear demarcations among these. We age in two ways: chronologically, the age derived from our birth certificates, and physiologically, the way in which our bodies actually function. Since these two ages do not always run in parallel, an individualized evaluation and treatment plan is essential.

Normal age-related changes must be distinguished from age-prevalent illness. Failure to do so will deny the older person benefit of treatment. In addition, numerous conditions present non-specifically or atypically in the elderly. Examples include myocardial infarction, acute abdomen, pneumococcal pneumonia, and hyperthyroidism. Changes in appetite, sleeping pattern, and mobility, including falls, may be the only early signs of some underlying problem. Unfortunately, the older person often "underreports" illness, wrongly assuming that disease states are part of the normal aging process. Since increases in morbidity and mortality may result from a delay in seeking medical assistance, patient education should be a major focus of geriatric health care.

Geriatric Profile

1. Normal age-related changes (universal, progressive, and irreversible)
2. Age-prevalent illness
3. Atypical and nonspecific presentation of disease

II. MEDICAL ASSESSMENT OF THE ELDERLY PATIENT

Comprehensive assessment of the older person must include a thorough evaluation of physical, mental, and social conditions. Although many physiological functions deteriorate with age, rarely does the age-related change give rise to any significant symptoms of pathology. "Normal" changes of aging, paucity of symptoms and signs, difficulty in history taking and examination, and atypical presentation of illness collectively contribute to the diagnostic difficulties commonly encountered by physicians when dealing with the diseases of old age. "Economy of diagnosis" is not the rule in geriatric medicine. It has been increasingly recognized in recent years that early and correct diagnosis is of vital importance in the successful and lasting treatment of the diseases of the elderly. A thorough understanding of physiological accompaniments associated with normal aging is essential. Misdiagnosis may not only affect the patient but often has serious implications for various other

members of the family. A summary of normal age-related changes
is listed.

III. Age-Related Physiological Changes

System	Effect of age	Consequences
Central nervous system	Decline in the number of neurons and the weight of the brain Reduced short-term memory Takes longer to learn new information Slowing of reaction time	May result in problems with independent living
Spinal cord/peripheral nerves	Decline in nerve conduction velocity Diminished sensation Decline in the number of fibers in the nerve trunks	Slowness of "righting" reflexes Muscle wasting Diminished sensory awareness Reduced vibratory sensation
Cardiovascular system	Reduced cardiac output (normal?) Valvular sclerosis of the aortic valves common Reduced ability to increase the heart rate in response to exercise	Reduced exercise tolerance
Respiratory system	Decline in vital capacity Increased lung compliance Reduced ciliary action Increased residual volume Increased AP chest diameter	Diminished oxygen uptake during exercise Reduced pulmonary ventilation on exercise Increased risk of pulmonary infection Reduced exercise tolerance
Gastrointestinal tract	Decrease in number of taste buds Loss of dentition Reduced gastric acid secretion Reduced motility of large intestine	Reduced taste sensation Possible difficulty in mastication Potential cause of iron deficiency anemia Constipation
Kidneys	Loss of nephrons Reduced glomerular filtration rate, and tubular reabsorption Change in renal threshold	Decreased creatinine clearance Reduced renal reserve may lead to reduced glycosuria in the presence of diabetes mellitus
Musculoskeletal system	Osteoarthritis Loss of bone density (normal?) Diminished lean muscle mass	Poor mobility; pain Decreased vertical height May predispose to fractures Change in posture Reduced strength
Endocrine/metabolism	Reduced basal metabolic rate (related to reduced muscle mass) Reduced glucose tolerance	Reduced caloric requirements Must distinguish from true diabetes mellitus
Reproductive	Men: Delayed penile erection, infrequent orgasm, increased refractory period, decreased sperm motility and altered morphology	Dimished sexual response Decreased reproductive capacity

(continued)

III. Age-Related Physiological Changes (*Continued*)

System	Effect of age	Consequences
	Women: Decreased vasoconges-tion, delayed vaginal lubrication, diminished orgasm, ovarian atrophy	
Skin	Loss of elastic tissue	Increased wrinkling; senile purpura
	Atrophy of sweat glands	Difficulty in assessing dehydration
	Hair loss	Reduced sweating
Sensory		
a. Eye	Arcus senilis	
	Lenticular opacity	Increased risk of falls and fracture
	Decreased pupillary size	Poor vision
	Contraction of visual fields	Presbyopia
b. Ears	Atrophy of external auditory meatus	
	Atrophy of cochlear hair cells	Presbycusis (loss of hearing of high frequencies)
	Atrophy of ossicles	Gradual loss of hearing
c. Taste	Reduced number of taste buds	Loss of interest in food
	Poor taste sensation	Malnutrition and weight loss
d. Smell	Decline in the sensation of smell	Increased risk of gas poisoning; decreased appetite

IV. AGE-PREVALENT ILLNESS

Certain conditions, such as Parkinson's disease, Paget's disease, dementia, osteoarthritis, osteomalacia, poor mobility, falls, and incontinence are seen far more frequently in old age. A summary of age prevalent illness is listed.

Age-Prevalent Disease

System	Disease
Central nervous system	Dementia
	Depression
	Parkinsonism
	Subdural hematoma
	Transient ischemic attack
	Trigeminal neuralgia
Eyes	Poor vision (cataract, macular degeration)
Ears	Poor hearing

(*continued*)

Age-Prevalent Disease (*Continued*)

System	Disease
Cardiovascular system	Hypertension Ischemic heart disease Arrhythmia Cardiac failure Peripheral vascular disease Varicose veins
Respiratory system	Chronic obstructive pulmonary disease Pneumonia Pulmonary tuberculosis
Endocrine/metabolic	Diabetes mellitus Hypothyroidism Hypokalemia/hyponatremia Gout
Gastrointestinal tract	Hiatus hernia Dysphagia Constipation Fecal incontinence Diarrhea Malabsorption syndrome Ischemic colitis Irritable bowel syndrome Rectal prolapse Carcinoma of colon
Genitourinary system	Urinary tract infection Urinary incontinence Prostatism Renal insufficiency Prostatic carcinoma
Musculoskeletal system	Osteoporosis Osteoarthrosis Osteomalacia Polymyalgia rheumatica Paget's disease of bone
Hematological system	Anemia Multiple myeloma Myelofibrosis
Autonomic nervous system	Hypothermia Postural hypotension
Oral pharanx	Edentulous; periodontal disease

(*continued*)

Age-Prevalent Disease (*Continued*)

System	Disease
Miscellaneous	Dehydration
	Foot problems
	Fractures
	Immobility
	Iatrogenic illness
	Malnutrition
	Cancer
	Pressure sores

V. ATYPICAL PRESENTATION OF DISEASE IN THE ELDERLY

Many diseases present atypically and non-specifically during later life. Examples of atypical presentation of illness are listed.

Examples of Atypical Presentation of Illness in Old Age

Disease	Examples of atypical presentation
Pneumonia	Anorexia
	Acute confusional state
	Normal pulse rate
	No elevation of body temperature
	No rise in white cell count
	Falls common
Pulmonary embolism	"Silent" embolism
	Nonspecific symptoms
Myocardial infarction	Anorexia
	Absence of chest pain
	General deterioration
	Falls
	Weakness
	Shortness of breath common
Congestive cardiac failure	Nonspecific symptoms
Acute abdomen	Absence of rigidity and tenderness

(*continued*)

Examples of Atypical Presentation of Illness in Old Age
(Continued)

Disease	Examples of atypical presentation
Urinary tract infection	Acute confusional state Absence of pyrexia No rise in white cell count Incontinence
Parkinsonism	General slowness Recurrent falls
Transient ischemic attacks	Acute confusional state Falls
Polymyalgia rheumatica	Nonspecific symptoms Poor general health Aches/pains Lethargy
Hyperthyroidism	Angina Atrial fibrillation Heart failure Absence of eye signs No increase in appetite Appetite may be poor Goiter commonly not palpated Bowel movements rarely increased
Hypothyroidism	Nonspecific deterioration Confusional state Depression Anemia Vaginal bleeding
Depression	May mimic dementia Weight loss
Malignancy	Nonspecific symptoms

VI. APPROACH TO THE ELDERLY PATIENT

Not uncommonly, the physician dealing with an elderly patient is confronted with difficulties in history taking and physical examination. Some suggestions as to how these difficulties may be overcome are listed.

Difficulties in History Taking in the Elderly

Problem	Suggestions
Poor memory; poor historical recall	Do not confuse poor attention with poor memory
	Seek history from close relatives, friends, and neighbors; obtain past medical records; seek the help of a social worker
	A number of diseases and/or medications may make the elderly patient confused, diminishing his/her ability to give a clear account of the events; carefully review and confirm history if necessary
Hearing impairment	Look for wax in the ear
	Speak slowly and clearly
	Face-to-face contact may help
	Do not rush the patient
	Make use of the speaking tube or ear trumpet (a stethoscope may be used)
	Try communicating through writing.
	Avoid shining light on patient's face (some elderly lip-read)
Speech problems	Seek the help of a speech therapist
	Edentulous patients may be difficult to understand, and dentures may help speech
Patient may not be forthcoming about some symptoms and may erroneously attribute symptoms to the normal aging process	Emphasize the value of giving a thorough history
	Ask questions regarding sleep, eating, sexual, bowel, and urinary function

VII. THE ELDERLY AND THE NURSING HOME

Approximately 5% of elderly persons (over age 65) live in institutional settings. Broken down by decade, this represents 2.3% of those over 65, 6% of those over 75, and 22% of those over 85. It is estimated that persons over the age of 65 have between a 25% and 40% chance of spending some time during their remaining life in a nursing home. Most elderly prefer the autonomy and comfort of their own homes and neighborhood and resist dependency or uprooting as long as possible. Once all efforts to keep an older person at home have failed, however, institutional care may be required. Home care programs, day care programs, day hospitals, and respite programs should be considered before recommending nursing home placement.

Leading causes of institutionalization include dementia, immobilization, and incontinence. Families must be supported during the decision-making process; since everyone has his own tolerance limit, the physician's own values must not enter into the decision process.

Common Characteristics of Nursing Home Residents

Mean age over 85
More women than men (2 : 1)
More whites than nonwhites
Multiple pathology
Most common problems are dementia, urinary incontinence, and poor mobility
Frequently, previous social support has been poor

The quality of care provided in the nursing home varies considerably. Some of the principles of good quality care are listed.

Principles of Good Quality Care in a Nursing Home

Independence must be encouraged
Family member interest and involvement must be maintained; relatives should be encouraged to take the resident home as often as they can; this may help minimize family guilt
The resident should be involved in an active program of rehabilitation and activities
Outdoor tours, trips, and activities should be arranged periodically
Privacy and dignity must be maintained
A relaxed environment, similar to a homelike setting, must be created

---

VIII. THE AGING PROCESS: THEORETICAL CONSIDERATIONS

There are many definitions of "aging." Perhaps the most comprehensive is a "progressive loss and deterioration of physiological capabilities and function resulting in decreased viability and increased vulnerability and probability of death." Why aging processes are "universal, progressive, and irreversible" remains a topic of great controversy. Though life expectancy has generally improved over the years, the life span or maximal period of survival has remained constant (approximately 110–120 years) since ancient times. Several theories of aging are listed.

Collagen Theory. Aging is associated with an increase in the cross linkages of the collagen fiber. Age changes result when two or more macromolecules become linked, and the molecules change their character or function. Collagen tissue gradually loses its elasticity, resulting in observed changes in skin, subcutaneous tissue, arteries, and joints.

Immune Function. It is well recognized that increasing age is associated with a decline in the body's immune system functioning. A decline in the discriminatory ability of the immune system may result in an increased incidence of autoimmune disorders and has been implicated in the pathogenesis of age-prevalent tumors. Various autoantibodies are increasingly seen in the otherwise healthy elderly individual. According to this theory, thymic involution with decreasing T-cell population and increasing autoantibodies are responsible for aging and various age-related diseases.

Finite Cell Life. Data suggest a maximum mitotic ability for various cells and species. This is thought by some to determine the life span of a species. Cells obtained from older persons and shorter-lived species are reported to have less potential for population doubling than cells obtained from younger persons and longer-lived species. This theory has been referred to as a "theory under glass," as experiments utilize tissue culture techniques.

"Hit" Theory. This theory proposes that longevity is determined by the cell's ability to repair damage that may occur to DNA. When UV light was used as a means to

(continued)

damage DNA, cells obtained from longer-lived species
and younger persons had the best repair capability.
Once again, this has been referred to as a "theory
under glass."

Free-Radical Theory. Scavenger cells are thought to circu-
late throughout the body, affecting cellular electrons
that determine whether a cell is in a "reduced" or
"oxidized" state. Antioxidants are thought to be capable of eliminating these scavenger cells and retarding
cellular aging.

Hormonal Theory. This theory is based on the belief that
various neural and hormonal functions control the process of aging and eventual death. The gradual decline
in hormonal function or in the cellular response to
hormones is thought to contribute to aging. Denckla
has postulated the presence of a new pituitary hormone
(DECO or decreasing oxygen consumption hormone)
that he feels regulates cellular aging. Tissue responsiveness to various hormones has also been shown to
be modified by age.

IX. MEDICATION USE IN THE ELDERLY

Although those over 65 years of age comprise only 11% of the
U.S. population, they utilize 25% of prescription and an even
higher percentage of over-the-counter medications.

Because of changes in life style, body physiology and composition, and more common multiple drug use, the elderly are
particularly prone to medication-related complications.

Major Problems Encountered with Medication
Use in the Elderly

1. Reduced compliance
2. Drug–drug interactions
3. Drug–nutrient interactions
4. Side effects often poorly tolerated
5. Paradoxical response may result from a given
   medication

X. ALTERATION IN PHARMACOKINETICS WITH AGE

Medications must be absorbed, distributed, metabolized, and cleared from the body. Even if medications are taken exactly as prescribed, age-related changes in body physiology and composition may result in altered handling of the medication, potentially increasing the risk of unwanted side effects or lack of efficacy. In addition, age-prevalent illness may further comprise medication utilization. (See table on pp. 14–15.)

XI. PRINCIPLES OF MEDICATION USE IN THE ELDERLY

1. Individualize medication dosage. "Start low, go slow." Allow equilibration to occur before increasing dosage; 3.3 serum half-lives usually are required before equilibration is reached.
2. Determine blood level whenever available. Peak and trough levels may be more necessary in treating the older patient. In addition, not all elderly can tolerate "therapeutic levels" (i.e., replacement thyroid hormone in an elderly hypothyroid individual with severe cardiovascular disease).
3. "Bioassay." Are you getting the clinical effect you want?
4. Carefully consider efficacy versus side effects.
5. Side effects well tolerated during youth may not be tolerated during older age (i.e., orthostatic hypotension has greater potential for inducing stroke, heart attack, and/or renal failure because of more common predisposing factors and reduced homeostatic responses). Avoidance is best treatment.
6. Always review all medications being taken, whether prescribed or over the counter. Medication interactions and side effects become additive and even synergistic as the number of medications being taken increases.
7. Limit the number of medications taken whenever possible. Compliance, interactions, and side effects are directly related to the number of medications taken.
8. Choose the form of medication most easily administered, i.e., pill, capsule, syrup, or injectable.

Normal Age-Related Changes Capable of Altering Pharmacokinetics

Function	Physiological change	Comments
Absorption	1. Reduced splanchnic blood flow	Reduced ability to absorb medication taken by oral route. In many elderly this is balanced by a decrease in bowel motility, allowing increased time for absorption to occur.
	2. Reduced parietal cell functioning	Less HCl acid production with age can alter "form" of medication to that less or more absorbable; i.e., dietary iron, ferric, must be converted at acid pH to absorbable form, ferrous; enteric coated aspirin may be broken down earlier than desired.
	3. Reduced cell number	Changes in cell number may result in more drug per unit cell. This is a particular problem with neuroleptic medications, often leading to exaggerated and even paradoxical responses.

Distribution	1. Changes in body composition	Increases in body fat with age may lead to excessive storage of fat-soluble medications (i.e., vitamin D); reduced ICF and ECF may lead to a reduced volume of distribution of water-soluble medications (i.e., alcohol), leaving more available for toxic effects.
	2. Thickening of basement membranes	May impair permeability of medication into cells.
	3. Reduced blood flow	May reduce ability of medication to reach desired site of action: suboptimal effect may occur despite a therapeutic blood level.
Metabolism and clearance	1. Decline in renal function	Renal function declines 0.6% per year after maturity. Potential for drug overdosing is ever present.
	2. Decreased cell numbers	May reduce metabolism of medication at cellular level.
	3. Subtle changes in liver function	Although rarely of clinical significance in persons with normal liver function tests, changes may result as a function of age, i.e., reduced first-pass clearance in the liver of propranolol.

9. Beware of child-proof containers. Use these only if truly indicated. Elderly persons with vision and arthritic problems have increased difficulty in opening these bottles.
10. All medications should be clearly labeled. Gluing pills on clearly labeled display cards with instructions for taking may be useful.
11. Instruct the patient what to do if a medication is inadvertently not taken.
12. Instruct the patient what to expect from a medication and what to do if the desired effect is not achieved.
13. Report all side effects promptly.
14. Keep dosage schedules as simple as possible.

SUGGESTED READING

Butler R. N., Lewis M. I.: Healthy successful old age, in: *Aging and Mental Health—Positive and Psychosocial Approaches*. St. Louis, C. V. Mosby, 1977, pp. 17–33.

Facts About Older Americans, 1980, Washington, U.S. Dept. of Health and Human Services, Office of Human Development Services, Administration on Aging, HHS Publication No. (80-20006), 1979.

Gambert S. (ed): *Contemporary Geriatric Medicine* (Vol 1). New York, London, Plenum Medical, 1983.

Gambert S. (ed): *Contemporary Geriatric Medicine,* (Vol 2). New York, London, Plenum Medical, 1986.

Kane R. L., Kane R. A.: *Assessing the Elderly: A Practical Guide to Measurement,* Lexington, MA, Heath, 1981.

Libow L. S., Sherman F. T.: Interviewing and history taking, in Libow L. S., Sherman F. T. (eds): *The Core of Geriatric Medicine* St. Louis, Mosby, 1981, pp. 1–4.

Rowe J. W. : Clinical research on aging: Strategies and directions. *N Engl J Med* 297:1332–1336, 1977.

2

Aging and the Cardiovascular System

I. INTRODUCTION

Cardiovascular disease is the leading cause of morbidity and mortality in the elderly.

Risk Factors for Cardiovascular Disease in the Elderly

1. Aging
2. Positive family history (becomes less important with increasing age)
3. Cigarette smoking
4. Hypertension
5. Obesity greater than 20% of average weight for age
6. Hyperlipidemia
7. Diabetes mellitus

Because of the heterogeneity of our older population, changes in the cardiovascular system are best thought of as being a composite of normal age-related changes and diseases of increasing prevalence with age.

Age-Related Changes in the Cardiovascular System Including Normal Age- and Disease-Related Findings

Anatomic

1. Cardiac muscle: increased left ventricular wall thickness and increased cardiac mass
2. Valves: calcification of aortic valve common
3. Conduction system: decrease in the number of sinoatrial node pacemaker cells; decreased cells within the bundle of His; amyloid infiltration in the bundle of His common
4. Arterial system: thickening of the medial layer of medium-sized arteries; atherosclerotic changes almost universally prevalent in members of industrialized societies

Physiological

At Rest

1. Increase in basal systolic blood pressure
2. Increase in basal diastolic blood pressure until seventh decade
3. Increase in pulse wave velocity
4. Decrease in early diastolic filling
5. Prolonged contraction phase
6. Increase in force of atrial contraction
7. Increase in afterload (impedence)
8. Increase in aortic volume
9. No change in cardiac output in healthy elderly

With Exercise

1. Decrease in overall maximum aerobic capacity
2. Decrease in maximally obtainable heart rate
3. Decrease in maximally obtainable ejection fraction
4. Increase in end-systolic volume
5. Increase in stroke volume

II. EVALUATION

The cardiac examination may be complicated by certain physical changes that often accompany advancing age. An increase in the anterior–posterior diameter of the chest wall may diminish the physician's ability to auscultate the heart sounds clearly. In addition, approximately 50% of elderly persons have at least one abnormal finding on routine electrocardiogram; in most cases these findings have no clinical significance.

Clinical Findings Often Noted during Routine Cardiac Examination of the Elderly

Physical Findings

1. Indistinct pulmonic component of S2. Auscultation over the carotid arteries should be done to rule out significant aortic stenosis. The presence of a preserved A2 component of the second heart sound decreases the probability of significant aortic stenosis.
2. Soft S4. This may be clinically significant if a palpable A wave is also noted.
3. Murmurs. Fifty to sixty percent of elderly persons have a clinically distinguishable murmur; the majority of these result from the ejection of blood into the aortic root.

Electrocardiographic Findings Commonly Noted in the Elderly

1. Left axis deviation
2. Primary atrioventricular heart block
3. Nonspecific ST–T wave changes
4. Left bundle branch block
5. Right bundle branch block

Although all of the above electrocardiographic findings are nonspecific, significance of the findings must be correlated with clinical history and the presence or absence of detectable heart disease.

III. ATHEROSCLEROTIC HEART DISEASE: ANGINA PECTORIS

Angina often manifests atypically in the elderly. Although chest pain is often a presenting finding, its presence or absence is not diagnostic. Angina may present with symptoms as diverse as dyspnea, lethargy, wheezing, and/or mental confusion. In general, the evaluation and treatment of angina should not be modified for age. It is essential, however, to choose carefully those medications that have a low side-effect profile with particular attention given to avoid volume depletion, electrolyte imbalance, and orthostatic hypotension. When patients present with symptoms of angina, careful evaluation of underlying risk factors is indicated.

Risk Factors for Angina Pectoris

1. Obesity
2. Previous myocardial infarction
3. Aortic valvular stenosis or regurgitation
4. Hypertension
5. Congestive heart failure
6. Hyperthyroidism
7. Anemia
8. Hyperlipidemia
9. Tachyarrhythmia
10. Cigarette smoking
11. Emotional and/or physical stress
12. Medications that increase myocardial oxygen consumption

The mainstay of treatment for angina pectoris includes medication such as the nitrates, β-blockers, and the newer calcium channel blockers. All have different mechanisms of action and act to reduce myocardial oxygen requirement and to increase coronary artery blood flow.

A. Medical Therapy of Angina Pectoris

1. Nitrates

Medication	Route	Dose	Excretion	Duration	Comment
1. Nitroglycerin	i.v.	50 μg/min	Hepatic	Rapid	Continuous drip for control of unstable angina or pain during MI
2. Nitroglycerin	s.l.	0.2–0.4 mg	Hepatic	30 min	Useful in acute angina; should cause burning sensation in mouth if preparation is active
3. 2% nitroglycerin ointment	Topical	0.5–0.3 in	Hepatic	6 hr	Preparation is systemically absorbed
4. Transdermal nitroglycerin	Topical	5–10 mg	Hepatic	24 hr	Long-acting patches may not have the desired hemodynamic response for the entire 24-hr period of application
5. Isosorbide dinitrate	p.o.	10–30 mg	Hepatic	4 hr	Reliable and effective; compliance is a major problem

Side effects for all of the nitrate preparations are similar. The major side effects are hypotension, headache, tachycardia, and nausea/vomiting. These symptoms may be more pronounced in the elderly and may serve to reduce compliance.

2. β-Blockade Therapy of Angina in the Elderly

Medication	Dose	Clearance	Half-life	Comment
1. Atenolol (Tenormin®)	50 mg t.i.d.	Renal	7 hr	Cardioselective; low lipid solubility
2. Metoprolol (Lopressor®)	50–200 mg q.i.d.	Hepatic	4 hr	Cardioselective; moderate lipid solubility
3. Nadolol (Corgard®)	40–320 mg q.i.d.	Renal	36 hr	Low lipid solubility; long-acting, once-daily dose often used
4. Pindolol (Visken®)	5–20 mg t.i.d.	Hepatic, renal	3.5 hr	Moderate lipid solubility
5. Propanolol (Inderal®)	20–40 mg q.i.d.	Hepatic	4.5 hr	High lipid solubility; proven efficacy in treatment after myocardial infarction
6. Timolol (Blocadren®)	10–30 mg b.i.d.	Hepatic	4.5 hr	Low lipid solubility; proven efficacy in treatment after myocardial infarction

β-blockers have proven efficacy in a number of medical conditions, including angina pectoris, supraventricular dysrhythmias, hypertension, postmyocardial infarction (decreases mortality), and hypertrophic cardiomyopathy. Caution must be exercised with the use of these medications in the elderly because of the many contraindications to β-blockade therapy and the side effect profile of these medications.

Contraindications to β-Blockade Therapy

1. Congestive heart failure except idiopathic hypertrophic subaortic stenosis
2. Bradycardia unless rhythm is paced
3. Asthma/chronic obstructive pulmonary disease
4. Diabetes mellitus
5. Hypotension

Side Effects of β-Blockade Therapy

1. Bronchospasm
2. Hypotension
3. Congestive heart failure
4. Bradycardia
5. Atrioventricular block
6. Peripheral vasoconstriction
7. Raynaud's phenomenon
8. CNS disturbances, depression, hallucinations, delirium
9. Impotence
10. Hypoglycemia
11. "Rebound"; must be weaned during cessation of therapy

3. Calcium Channel Blockade Therapy in the Elderly

Medication	Route	Dose	Clearance	Half-life	Comment
1. Verapamil (Calan®)	p.o., i.v.	80–120 mg t.i.d.	Hepatic	3–7 hr	Decreases atrioventricular conduction; decreases myocardial contractility; useful in many supraventricular tachyarrhythmias
2. Nifedipine (Procardia®)	p.o., i.v., s.l.	10–40 mg q.i.d.	Hepatic	4–5 hr	May cause tachycardia if autonomic nervous system is intact
3. Diltiazem (Cardiazem®)	p.o.	30–60 mg q.i.d.	Hepatic	4–6 hr	Lowest incidence of side effects of the calcium channel blockers

The calcium channel blockers have relatively few side effects compared to the other antianginal agents. Verapamil should not be used in patients with left ventricular dysfunction or in diseases of the atrioventricular nodal system. Nifedipine may produce dilatation of peripheral arteries; this may result in reflex sympathetic activity causing a sinus tachycardia that may serve to aggravate angina by increasing myocardial oxygen consumption. The peripheral vasodilatation may also induce flushing, severe headache, and hypotension. Finally, few side effects are seen with the appropriate use of diltiazem at clinically effective doses.

B. Summary

The goal of therapy for angina is to maintain a functionally independent life style free of signs and symptoms. Therapy should start with the nitrates and progress to the β-blockers. There is good evidence that the two agents may have an additive effect in the control of symptoms. Care must be taken to choose an appropriate calcium channel blocker as a third agent. Diltiazem may be safely used with both nitrates and β-blockers. When the patient is symptom-free on the prescribed medication regimen, a low-level stress test with nuclear scan to test myocardial perfusion is suggested. If results still indicate a decrease in myocardial perfusion during exercise, coronary angiography may be indicated. In all cases, the possible benefit of the diagnostic evaluation and treatment offered must clearly outweigh potentially harmful side effects. If the diagnosis is variant angina (Prinzmetal's angina), coronary vasospasm may respond more rapidly to the calcium channel blockers, the drug of choice for this condition.

IV. MYOCARDIAL INFARCTION

In general, the same guidelines for therapy of myocardial infarction (MI) should be followed regardless of age. Consideration must be given, however, to dosages of medications used for elderly patients. Lidocaine HCl, for example, when given in excessive dosage as a prophylaxis against MI-associated arrhythmias, can result in significant central nervous system side effects. These include seizures and delirium and may contribute to increased morbidity and mortality in the elderly patient. The adage of "start low, go slow" and close monitoring must be followed for all aspects of treatment.

Common Post-MI Complications in the Elderly			
Complication	Risk factors	Location of MI	Presentation
1. Ventricular septal defect	Age; hypertension; left ventricular hypertrophy; women	Anterior wall; posterior wall	Biventricular failure; 3–5 days post-MI; harsh holosystolic murmur
2. Rupture of left ventricular wall	Age; first MI	Anterior wall	Recurrent chest pain; sudden shock; terminal arrhythmia; 3–6 days post-MI
3. Papillary muscle rupture	Age	Posterior–inferior wall; anterior wall	Left ventricular failure; hypotension; pulmonary edema; 3–5 days post-MI

V. CONGESTIVE HEART FAILURE

Approximately 75% of patients suffering from congestive heart failure are over 60 years of age.

Common Causes of Congestive Heart Failure in the Elderly

Cause	Comment
1. Myocardial ischemia	Usually secondary to previous myocardial infarct; prognosis dependent on left ventricular wall motion and function
2. Hypertension	Increased work load because of long-term elevation in peripheral vascular disease leads to left ventricular failure
3. Calcific degeneration of a cardiac valve	Aortic stenosis is most common in the elderly
4. Amyloidosis	Clinically resembles constrictive pericarditis with decreased QRS voltage on ECG; ventricular conduction defects
5. Hypertrophic cardiomyopathy	Paradoxical response to conventional therapy; diagnosis usually made on echocardiogram
6. Miscellaneous	Thyroid disease; anemia; pulmonary embolism; infection; arrhythmias; alcohol toxicity

The evaluation of congestive heart failure must exclude any immediately treatable causes. In most cases the proper diagnosis will direct therapy. Previously it was felt that mitral valve prolapse and hypertrophic cardiomyopathy were uncommon causes of congestive heart failure in the elderly; recent evidence, however, indicates a much higher prevalence in this population. Physical findings of congestive heart failure, i.e., dyspnea, peripheral edema, jugular venous distention, hepatic congestion, etc., although clinically useful to assess the severity of the problem, are usually not helpful in the determination of the underlying etiology. Therefore, further laboratory evaluation is usually indicated. One of the most important diagnostic tests is the echocardiogram to assess wall motion, wall thickness, and valvular function.

Medical therapy is aimed at reducing the work load on the heart and making the heart pump more efficiently. Therefore, therapy is directed at decreasing preload (end-diastolic cardiac volume) and afterload (impedance) and increasing myocardial contractility. (See table on p. 25.)

Medical Therapy of Congestive Heart Failure in the Elderly

Medication	Mechanism of action	Average daily dose	Clearance	Duration of action	Comments
1. Digoxin	Increases myocardial contracility	0.125 mg	Renal	36 hr	Inotropic action may not outweigh side effects in elderly; following serum levels usually not clinically useful
2. Hydrochlorothiazide	Preload reduction; diuretic	25–50 mg	Renal	12 hr	Site of action at distal tubules
3. Furosemide (Lasix®)	Preload reduction; diuretic	40–120 mg	Renal	6–8 hr	Site of action at loop of Henle
4. Nitrates	See antianginal therapy; preload reduction				
5. Hydralazine	Afterload reduction; vasodilatation	50–200 mg	Renal	6–8 hr	May cause orthostatic hypotension
6. Nifedipine	See antianginal therapy; afterload reduction				
7. Prazosin	Preload and afterload reduction; blocks vascular α-adrenergic receptors	3–20 mg	Hepatic	4–8 hr	May get increased fluid retention; "first-dose" phenomenon common; may become tolerant to therapy
8. Captopril	Preload and afterload reduction; angiotensin-converting enzyme inhibition	75–300 mg	Renal	4–6 hr	May cause depressed renal function and neutropenia

VI. VALVULAR HEART DISEASE

Evaluation of the cardiac valves in the elderly requires careful consideration of the differential diagnoses and a thorough physical evaluation.

Differential Diagnosis of a Systolic Murmur in the Elderly

1. Aortic stenosis
2. Aortic sclerosis
3. Mitral regurgitation
4. Papillary muscle dysfunction
5. Mitral annular calcification
6. Idiopathic hypertrophic sub-aortic stenosis

Hemodynamic Maneuvers That Are Clinically Useful to Evaluate Systolic Murmurs in the Elderly

	Intensity of murmur			
Condition	Standing upright	Squatting	Valsalva	Isometric
1. Aortic stenosis	Decrease	No change	Decrease	Decrease
2. Hypertrophic cardio-myopathy	Increase	Decrease	Increase	Decrease
3. Mitral valve prolapse	Increase	Decrease	Variable	Variable
4. Mitral re-gurgitation	No change	Increase	Decrease	Increase

---

A. Aortic Stenosis

Aortic stenosis is a common valvular problem in the elderly. Careful evaluation must be undertaken early to detect signs and symptoms because of the need for surgical replacement of the seriously affected valve.

Common Etiologies of Aortic Stenosis in the Elderly

1. Degenerative calcification of aortic valve—most common cause of aortic stenosis in patients over 65 years old.
2. Calcified bicuspid valve—most common cause of aortic stenosis in patients under the age of 65 years but still seen in the elderly.
3. Rheumatic heart disease—rarely seen in elderly without concomitant mitral valve disease

In evaluating for aortic stenosis, clinical signs often help to differentiate from aortic sclerosis.

Findings That Favor the Diagnosis of Aortic Stenosis

1. Loss of A2 component of the second heart sound; must be auscultated over carotid arteries
2. Palpable systolic thrill
3. "Ring calcification" on chest radiograph
4. Increased ventricular wall thickness and decreased motion on echocardiogram
5. Doppler echocardiography correlates well with the aortic valve gradients found on coronary angiography; this finding is less helpful if aortic regurgitation is present
6. Coronary angiography is the diagnostic test of choice

When aortic stenosis becomes symptomatic, mortality is 48% at 2 years and 64% at 5 years. These figures vary according to the symptoms.

Mortality Associated with Symptoms of Aortic Stenosis

Symptoms	Life expectancy
1. Syncope	3–4 years
2. Angina pectoris	2–3 years
3. Heart failure	1–2 years

Even after the appearance of these symptoms, surgical replacement of the affected valve is indicated and should be considered as early as possible. Successful surgical intervention has been shown to increase life expectancy considerably.

B. Aortic Regurgitation

Aortic regurgitation (AR) in the elderly can lead to an overload of the left ventricle and clinical signs and symptoms of congestive heart failure. Because of the decreased distensibility of the aorta in the elderly, the wide pulse pressure associated with aortic regurgitation in a younger population is often not found. In addition, the diastolic murmur may be difficult to distinguish in this population.

Common Causes of Aortic Regurgitation in the Elderly

1. Rheumatic heart disease: usually seen in conjunction with mitral valve disease
2. Infectious endocarditis: most common cause of AR in the elderly
3. Congenitally bicuspid valve: usually seen in younger population
4. Trauma
5. Collagen disease: rheumatoid arthritis, systemic lupus erythematosus, etc.

C. Mitral Stenosis

Mitral stenosis is a relatively uncommon finding in the elderly. The presentation may be insidious; dyspnea on exertion and orthopnea are difficult symptoms to evaluate.

Possible Causes of Mitral Stenosis in the Elderly

1. Rheumatic heart disease   3. Bacterial endocarditis
2. Left atrial myxoma

Therapy is usually medical until the patients progress to a functional class II-B (symptomatic with moderate activity despite therapy). Surgical replacement of the mitral valve may be indicated at this point; consideration of the patient's overall physical condition is necessary prior to surgical intervention.

D. Mitral Regurgitation

As with the other valvular diseases in the elderly, mitral regurgitation is commonly seen.

Etiologies of Mitral Regurgitation

Acute
1. Rupture of chordae tendonae
2. Rupture of papillary muscle
3. Perforation of a mitral valve leaflet

Chronic
1. Rheumatic heart disease
2. Mitral valve prolapse
3. Coronary artery disease
4. Progressive dilation of left ventricle
5. Calcification of the mitral annulus
6. Collagen tissue disorder

Of the etiologies listed above, calcification of the mitral annulus is the most common cause of mitral regurgitation in the elderly.

Previously mitral valve prolapse was thought to be rare in the elderly. Recently, however, an increased prevalence has been described.

The presence of mitral valve prolapse in the elderly is associated with a worse prognosis than that during youth. The natural history of this problem in the elderly, however, remains unknown.

Therapeutic Considerations of Mitral Valve Prolapse in the Elderly

1. Prophylactic antibiotics should be used prior to dental and surgical procedures
2. Evaluate the need for antiarrhythmic therapy; the incidence of sudden death is probably higher in the elderly

---

(continued)

**Therapeutic Considerations of Mitral Valve Prolapse
in the Elderly** (*Continued*)

3. Underlying congestive heart failure must be treated
   early; surgical replacement of the valve may be
   necessary.

VII. INFECTIVE ENDOCARDITIS

The presence of aortic valvular degeneration or calcification of
the mitral annulus predisposes the elderly individual to a relatively
high incidence of infective endocarditis.

Common Pathogens Seen in Bacterial Endocarditis

1. *Streptococcus viridans:* most commonly seen in elderly persons following dental procedures
2. *Streptococcus bovis:* commonly associated with tumor
   of the gastrointestinal tract
3. *Streptococcus faecalis:* associated with genitourinary
   tract infections or manipulation
4. *Staphylococcus aureus:* associated with surgical procedures, particularly heart valve replacement
5. *Staphylococcus epidermidis:* associated primarily with
   cutaneous infections
6. Gram-negative organisms: associated with gastrointestinal tract manipulation
7. Fungus: associated with immunologic deficiencies and
   surgical valve replacement

Treatment should be directed at the specific infecting organism
following blood cultures.

VIII. ARRHYTHMIAS

Conduction disturbances constitute one of the major cardiovascular health problems in the elderly. Many arrhythmias may
be secondary to age-related changes that occur in the heart.
Symptoms associated with arrhythmias may range from a mild
"lightheaded" feeling with occasional palpitations to syncope and
sudden death. In every case, careful evaluation and timely intervention are potentially life saving.

Classification of Tachyarrhythmias in the Elderly

Rhythm	Treatment
1. Sinus tachycardia	Reverse underlying etiology, i.e., volume depletion; maintain hemodynamic stability; prolonged rapid rate can precipitate myocardial infarction
2. Paroxysmal supraventricular tachycardia	Digoxin is drug of choice; if patient is hemodynamically unstable, DC countershock is indicated; if unresponsive to medications or countershock, pacing of the right atrium may be indicated
3. Atrial flutter	Digoxin is drug of choice to reduce ventricular rate to less than 100 beats/min; hemodynamic instability requires DC cardioversion
4. Atrial fibrillation	Digoxin to reduce heart rate to less than 100 beats/min; usually less digoxin is required than used in atrial flutter; quinidine may be added cautiously after digitalization to attempt induction of chemical cardioversion; DC countershock is often effective, but atrial fibrillation will reoccur if atria are enlarged
5. Nonparoxysmal supraventricular tachycardia	Discontinue use of digoxin because of toxic effect; often unresponsive to other antiarrhythmic medications; DC countershock is contraindicated
6. Ventricular tachycardia	A medical emergency that requires immediate countershock; recurrent episodes must be treated with antiarrhythmic medications.

Classification of Bradyarrhythmias in the Elderly

Rhythm	Treatment
1. Sinus bradycardia	Commonly occurs in elderly; rarely requires treatment

(continued)

Classification of Bradyarrhythmias in the Elderly
(Continued)

Rhythm	Treatment
2. Sinus node impairment (sinoatrial exit block or sinus pauses)	Cardiac pacing only if symptomatic with hypotension and/or congestive heart failure
3. Complete heart block (atrioventricular dissociation)	Cardiac pacing required
4. Mobitz I atrioventricular block	Cardiac pacing almost always indicated
5. Variable heart block	
a. Narrow QRS complex	Usually means AV nodal disease; rarely requires treatment
b. Wide QRS complex	Problem is below the level of AV node; if defect localized to His bundle or ventricle, pacing may be indicated

A. Ventricular Arrhythmia

The presence of a ventricular rhythm disturbance in the elderly is often associated with an increased risk of sudden death. Although many elderly have occasional premature ventricular contractions (PVCs), the presence of symptoms, previously existing structural heart disease, and the level of PVC will influence the ultimate prognosis. Careful evaluation and therapy with an appropriate antiarrhythmic medication are indicated to decrease the risk of sudden death.

Factors Associated with High Risk of Sudden Death

1. History of palpitations, syncope, angina, congestive heart failure, previous episode of cardiac arrest
2. Structural heart disease or ventricular dysfunction, coronary heart disease, ventricular dilatation, ventricular hypertrophy
3. Level of premature ventricular contraction; frequency more than 30 per min is associated with worse prognosis (PVC); multiformity, multifocal PVCs; repetitiveness, couplet; ventricular tachycardia; timing, R-on-T phenomenon

---

Treatment of Ventricular Arrhythmias

1. Correct any underlying precipitating factors: electrolyte disorders; hypoxia; drug toxicity; heart failure
2. Establish base-line measurement of ventricular arrhythmia; obtain a 24-hr Holter tracing if possible prior to initiating therapy
3. Initiate therapy with antiarrhythmic agents
4. Monitor serum for therapeutic drug levels, readily available with quinidine and procainamide
5. Assess efficacy of treatment with 24-hr Holter monitoring and/or electrophysiological testing
6. Adjust dosages according to toxicity, serum levels, and results of monitoring

B. Commonly Used Antiarrhythmic Medications

Type I Antiarrhythmic Agents				
Medication	$T_{1/2}$	Metabolism	Dosage range	Side effects
Type IA (membrane stabilization; prolongs conduction)				
1. Quinidine sulfate	5–7 hr	Hepatic with some renal excretion	200–400 mg q6h p.o.	Hypotension; prolonged QT interval
Quinidine gluconate	5–7 hr	Some active metabolites	200–400 mg q6h p.o.	Heart block; nausea/vomiting; depression; hyperglycemia; diarrhea; increased serum digoxin levels
2. Disopyramide phosphate (Norpace®)	4–8 hr	40–60% renal excretion; hepatic degradation; no active metabolites	100–300 mg q6h p.o.	Prolonged QT interval; AV block; anticholinergic effects; myocardial depression; gastrointestinal symptoms
3. Procainamide HCl (Pronestyl®)	2–4 hr	Renal excretion of drug and active metabolite	One-gram loading dose in first 24 hr followed by 250–500 mg q3h p.o.	Prolonged QT interval; nausea/vomiting; lupus erythematosis; myocardial depression; diarrhea
Type IB (membrane stabilization; enhances conduction)				
1. Lidocaine HCl	100 min after tissue loading with i.v. dose	Hepatic with active metaboite	1–3 mg/min i.v. drip	Focal and grand mal seizures; respiratory arrest; dizziness; AV block; sinoatrial arrest; delirium

(continued)

Type I Antiarrhythmic Agents (*Continued*)				
Medication	$T_{1/2}$	Metabolism	Dosage range	Side effects
2. Phenytoin (Di-lantin®)	4–6 hr	Hydroxylated in liver; excreted in urine	1 g p.o. in divided doses over the first 24 hr, then 200–500 mg p.o. daily	Conduction in atrial flutter; respiratory arrest; idio-ventricular rhythm

IX. AORTIC ANEURYSMS

Approximately 40% of individuals over 80 years of age have some form of aortic aneurysm. Approximately 2% of the geriatric population will have a clinically detectable aneurysm, the majority (80%) occurring in the abdomen distal to the renal arteries. Diagnosis is usually made on routine physical examination and may be confirmed with either lateral lumbosacral spine X rays (the majority of aneurysms contain calcium deposits, making the lesion radiopaque) or ultrasonography. The exact size of the aneurysm should be confirmed, since the anterior–posterior diameter of the aneurysm is often the largest dimension. The size of the aneurysm is critical to the determination of the treatment. If the aneurysm is found to be 6 cm at its widest diameter, close observation is recommended; the natural history is one of gradual expansion, especially in the elderly. Sudden increases in size do occur, and the incidence of rupture even in aneurysms less than 6 cm in diameter is higher than had been previously thought. Careful consideration of the patient's overall medical condition and functional capabilities must be employed when deciding on a treatment plan. In most cases, aneurysms greater than 6 cm should be surgically repaired unless medically contraindicated. Unfortunately, surgical mortality for repair of abdominal aortic aneurysm is considerable, and so when to repair asymptomatic aortic aneurysms still remains an area of controversy within the medical literature. Time from diagnosis to rupture varies with age, with those under age 65 having a mean of 18 months versus 30 months for those over age 65. These "soft" retrospective data may persuade the clinician to continue medical management of even a large aneurysm in the frail, chronically ill elderly patient.

X. AORTIC DISSECTION

Aortic dissection is a medical emergency that most commonly occurs in men over fifty years of age. It is usually characterized by the sudden onset of excruciating chest pain radiating to the scapula region; the pain, however, may also be felt in the neck, jaw, arm, abdomen, and/or leg.

There are basically two major sites of aortic dissection, each of which is treated differently and has a different prognosis. The most common location for aortic dissection is the ascending aorta, representing approximately 70% of all cases. Presentation at this site is the most damaging and often has a poor outcome. The descending aorta is the site for approximately 30% of dissections. Although it is still a medical emergency, the prognosis is much better.

The diagnosis is often difficult, and a thorough evaluation must be done to rule out a variety of other problems.

Differential Diagnosis of
Aortic Dissection

1. Acute myocardial infarction
2. Pericarditis
3. Pulmonary embolus
4. Pneumothorax
5. Biliary colic
6. Acute cholecystitis
7. Appendicitis

Contrast arteriography remains the test of choice for making an accurate diagnosis. This test is both sensitive and specific. The computed tomography (CT) scan with contrast of the aorta can often highlight the false double aortic lumen seen in aortic dissection. This test, however, remains only an adjunct in diagnosis. Other modalities have recently shown promise in helping make the diagnosis. These include two-dimensional echocardiography and digital subtraction angiography. More work will be necessary before either of these may be substituted for contrast angiography.

Rapid diagnosis and treatment are essential for aortic dissection because of the high mortality associated with this problem. The mortality of untreated aortic dissection is 1% per hour for the first 24–48 hr after the occurrence. The initial treatment is largely medical: blood pressure must be lowered with antihypertensive medication (often sodium nitroprusside i.v. with an appropriate arterial line in place for monitoring purposes), and a negative inotropic agent should be given to reduce the work load of the heart and the systolic blood pressure (often propranolol). The goal of medical therapy is to keep the mean arterial pressure in the range of 60–70 mm Hg. In the case of the distal dissection, long-term medical management if successful is often preferred to surgery. As soon as possible, oral antihypertensive medication and oral β-blockers should be substituted for the i.v. medication used during the acute episode. In addition, there must be careful follow-up of signs and symptoms, and an annual CT scan of the aorta should be performed. This conservative management carries an in-hospital mortality rate of 20% during the acute episode.

In cases of proximal dissection of the ascending aorta, more aggressive management is required. Surgical repair of this lesion is the treatment of choice even though the mortality associated with surgery is close to 40%. Medical management of the same condition is associated with a mortality of 70–80%.

XI. CARDIAC RISKS DURING SURGERY

A thorough preoperative medical evaluation is essential in all elderly patients. Careful evaluation of cardiac risk factors is required to make an accurate assessment of surgical risk.

Factors That Influence Cardiac Risk during Noncardiac Surgery*

1. Presence of S3 gallop or jugular venous distension on preoperative physical examination — 11 Points
2. Recent (within 6 months) myocardial infarction — 10
3. Frequent premature ventricular contractions (> 5/min) — 7
4. Rhythm other than sinus — 7
5. Age over 70 years — 5
6. Emergency operation — 4
7. Intrathoracic, intraperitoneal, or aortic surgery — 3
8. Hemodynamically significant aortic stenosis — 3
9. Poor general medical condition — 3

---

*Adapted from Goldman, L.: Cardiac risks and complications of noncardiac surgery. *Ann Intern Med* 98:504–513, 1983.

From this point scale a relative cardiac risk index has been developed.

Relative Cardiac Risk Index for Noncardiac Surgery*

1. Class I   0–5 points; associated with a very low risk; in general, less than 0.5% mortality

2. Class II   6–12 points; associated with a 5% cardiac complication rate and a 2% mortality rate

3. Class III   13–26 points; associated with an 11% cardiac complication rate; mortality remains low at 2%

4. Class IV   26 points; associated with high cardiac morbidity (22%) percent) and very high cardiac mortality (56%)

---

*Adapted from Goldman, L.: Cardiac risks and complications of noncardiac surgery. *Ann Intern Med* 98:504–513, 1983.

More recently, the inability to increase the heart rate above 99 beats/min after 2 min of exercise has been shown to be associated with a higher incidence of perioperative cardiac complication; however, more work will be necessary to determine the importance of this finding.

Cardiac surgery must be viewed differently and still remains an area of controversy. Even though cardiac surgery carries a greater risk for the elderly, the risk/potential benefit ratio must be carefully evaluated in order to make the therapeutic choice. Age should never be used as the sole determinant of surgical risk. Other factors, such as general medical condition and presence or absence of preexisting severe cardiac disease, must be considered. Each case must be individualized. For example, untreated hemodynamically significant valvular aortic stenosis accompanied by symptoms has a high overall mortality rate. Even though the risk of perioperative complications may be high, surgical replacement of the aortic valve is often indicated. Coronary artery bypass surgery continues to remain an area of controversy. Recent data suggest that although the perioperative cardiac complication rate (5–10%) is higher in the elderly, continued medical management of left main or three-vessel coronary artery disease may result in an even higher mortality rate.

XII. ORTHOSTATIC HYPOTENSION

Orthostatic hypotension is a potentially life-threatening problem. It is defined as a decline in systolic blood pressure with a change to an upright position, with or without a reflex increase in pulse rate. The symptoms may range from a mild "lightheaded" feeling to a syncopal episode. Orthostatic hypotension is caused by either volume depletion or an impairment of the autonomic nervous system. If the autonomic nervous system is impaired, there will be no reflex rise in pulse rate accompanying the drop in blood pressure.

As previously mentioned, the reflex autonomic nervous system response is somewhat blunted in the elderly, and a significant volume depletion may not be diagnosed if one expects to see an increase in pulse rate.

Physiological Response to Standing

1. Decrease in venous return to the heart
2. Decrease in cardiac output
3. Decrease in arterial blood pressure
4. Activation of baroreflex response
5. Increase in sympathetic tone
6. Decrease in vagal tone
7. Increase in heart rate

There are many causes of either the significant volume depletion or the impairment of the autonomic nervous system in the elderly. The most common causes include dehydration and/or medications; however, the differential diagnosis is long and requires careful evaluation prior to the initiation of treatment.

Etiologies of Orthostatic Hypotension in the Elderly

1. Medications: antihypertensives; antidepressants; antipsychotics; antiparkinson agents; antiemetics; sedatives; hypnotics; analgesics; bronchodilators; diuretics
2. Neurological impairment of autonomic reflexes: idiopathic orthostatic hypotension; Parkinson's disease;

(continued)

Etiologies of Orthostatic Hypotension in the Elderly
(*Continued*)

aging; alcohol; peripheral neuropathies; Guillan-Barré syndrome; tabes dorsalis; cervical trauma; multiple sclerosis
3. Cardiac dysfunction: congestive heart failure; ventricular arrhythmias
4. Metabolic: dehydration; amyloidosis; diabetes mellitus; pheochromocytoma

---

Recently, a postprandial decline in systolic blood pressure has been described. This is more pronounced in the aged and may exacerbate already borderline changes in postural blood pressure. Patients should be advised to rise from the dining table slowly and not immediately after eating, thus preventing complications from this apparently physiological finding.

The treatment for orthostatic hypotension should be directed at the underlying etiology. The elderly patient should be advised to change positions slowly to allow for equilibration. In addition, the volume status of the patient should be evaluated thoroughly. If the underlying etiology is not related to a known reversible cause, careful use of medications such as fluorocortisone, 0.2–1.0 mg/day, may be initiated with careful titration.

Other medications that may be helpful include propranolol, 80–160 mg/day, or indomethacin, 25–50 mg t.i.d., if not otherwise contraindicated.

As with all medications in the elderly, the therapeutic benefit must outweigh the potential side effects. Careful monitoring is essential for quality care. Special positive-pressure pants and boots have also been suggested in select cases.

XIII. HYPERTENSION IN THE ELDERLY

Approximately 15 to 20% of all adults, and an even higher percentage of those over 65, have elevations in blood pressure, a leading cause of both mortality and morbidity in America.

According to the World Health Organization, regardless of age, normal blood pressure is defined as less than 140/90 mm Hg. Hypertension is any blood pressure in excess of 160/95 mm Hg. Systolic pressure increases gradually during childhood until the early adult years, when a more rapid change occurs. Diastolic blood pressure increases slowly until approximately 60 years of age with a tendency to decline or remain constant thereafter.

A. Epidemiology

The following criteria have been epidemiologically related to the prevalence of hypertension in the United States:

1. Age. The prevalence of high blood pressure increases with advancing age, affecting 40% of those over age 65.
2. Race. American blacks have the highest prevalence of hypertension for all aged cohorts of both men and women. Environment and genetic influences may be contributing factors.
3. Sex. Men are affected with hypertension more commonly below age 45; after menopause, however, women are affected with hypertension more than men.
4. Diet. Data suggest a direct relationship between sodium intake and blood pressure in a select "salt-sensitive" population. Alcohol intake also has been directly linked to hypertension. Potassium intake, on the other hand, appears to be protective. Other nutrients, including iron, cadmium, phosphorus, and calcium have also been linked to changes in blood pressure, although conclusive data are lacking.
5. Weight. Obese individuals are more likely to be hypertensive.

B. Pathophysiology

Physiological accompaniments of the "normal" aging process and diseases more prevalent with age may result in factors capable of either increasing or decreasing blood pressure.

Factors potentially capable of elevating blood pressure include collagen degeneration with sclerosis and fibrosis in many organ systems, decreased baroreceptor sensitivity, increased cardiac after-load, increased peripheral vascular resistance, increased nor-epinephrine levels, and decreased aortic distensibility.

Those factors potentially capable of reducing blood pressure with age include decreased myocardial contractility, ejection fraction, or cardiac output; increased aortic volume; decreased renin response; decreased number and sensitivity of adrenergic receptors.

The decrease in plasma renin levels associated with increasing age most likely results from a reduction in the responsiveness of the juxtaglomerulus apparatus. Levels of aldosterone are highest during infancy and gradually decline with advancing age. One possible etiology is a decline in endocrine function of the kidney as part of the normal aging process with thinning of the renal cortex and loss of renal mass. In addition, circulating catecholamines are less able to stimulate renin secretion with age. This is thought to be secondary to the decreased number and sensitivity of β-adrenergic receptors in the kidney.

Although it is impossible to predict which, if any, of the above factors will influence the blood pressure in a given individual, in the elderly, systolic blood pressure is more commonly increased compared to diastolic pressure. Isolated elevations in systolic blood pressure can be found in approximately 40% of hypertensive patients above the age of 65 as compared to only 10–15% of those hypertensive persons under the age of 40. This most likely results from age-related vascular alterations. Large arterial vessel walls become thick as subendothelial lipid deposits accumulate along with increasing amounts of collagen, glycosaminoglycans, and calcium. In addition, elastin fibers degenerate. The end result is an arterial system with impaired distensibility and decreased capacitance. As the ability to absorb pressure with systole diminishes, systolic pressure rises.

C. Secondary Causes of Hypertension in the Elderly

Ninety-five percent of all hypertension is considered essential or of unknown cause. Specific etiology for increased blood pressure (secondary hypertension), however, can be identified in approximately 5% of cases.

Causes of Secondary Hypertension

Acromegaly	Hyperparathyroidism
Adrenal cortical disorders	Hyperthyroidism
Anemia	Hypothyroidism
Anxiety	Pheochromocytoma
Central nervous system	Polycythemia
disorders	Renal parenchymal lesions
Estrogen therapy	Renal vascular lesions

In the elderly, parenchymal and vascular disorders of the kidney are the most common, with pressure elevations resulting from fluid and sodium retention as a consequence of diminished glomerular filtration and/or increased renin release. Although the most common cause of renovascular hypertension during youth is fibromuscular dysplasia, elderly more commonly present with acute blood pressure elevations as a result of atherosclerotic narrowing of renal blood vessels.

Since adrenocortical steroid hormones have sodium- and fluid-retaining properties similar to those of aldosterone, it is important to consider and rule out when likely a functioning adrenal carcinoma, ACTH-producing pituitary tumor (Cushing's disease), or ectopic ACTH-secreting tumor, all potential causes of hypertension. Several other endocrine disorders have also been associated with elevated blood pressure: excess aldosterone secretion caused by primary hyperaldosteronism; increased catecholamines resulting from a pheochromocytoma; and the hypermetabolic state of hyperthyroidism. In addition, although not well understood, hypertension has been associated with hypothyroidism and acromegaly. Increased blood pressure has also been described in hyperparathyroidism. This most likely results from a calcium-induced vasoconstriction. Estrogen therapy, particularly in the dosages used for oral contraception, also has been linked to hypertension. Although most elderly women on estrogen replacement are on low dosages, increased renin levels may also result. Other causes of increased blood pressure include anxiety with changes in the autonomic nervous system, severe anemia with a compensatory high cardiac output, polycythemia, and central nervous system disorders.

D. Evaluating the Hypertensive Patient

A comprehensive history and physical examination are essential. Emphasis must be placed on obtaining information that might suggest the presence of an etiologic factor. Target end-organ damage related to the hypertensive state (for example, retinal changes or signs of cardiac dysfunction) must also be carefully looked for. Because the yield is so low in this population, exhaustive evaluations for secondary causes of increased blood pressure should only be done if indicated by the history or physical examination or when a rapid elevation in blood pressure has occurred. Routine screening laboratory tests should include those that (1) might document evidence of an unsuspected etiology or major target organ involvement (chest X ray, electrocardiogram, renal function tests); (2) are necessary for base-line determinations prior to initiating therapy (complete blood count, liver function studies); and (3) can be directly affected by anticipated treatment (electrolytes).

Suggested Initial Workup for the Hypertensive Elderly

1. Accurate blood pressure readings three times
2. Complete history
3. Complete physical examination
4. Laboratory data to include urinalysis, complete blood count, SMA-20, chest roentgenogram, electrocardiogram

Pseudohypertension, a falsely high blood pressure reading because of poor vascular compressibility, must also be considered. Treatment in these cases could reduce blood pressure to a dangerously low level.

Prior to beginning treatment, it is essential to document elevations in blood pressure carefully. Attention must be given to the choice of an appropriately sized blood pressure cuff—either a pediatric cuff for small arms or a leg cuff for obese arms. Customarily, systolic and diastolic pressures are recorded on three separate visits. Because an auscultatory gap is not an uncommon finding in the elderly, measurement technique is critical. The right brachial artery should be palpated as the sphygmomanometer pressure is raised by 10-mm Hg increments. At the disappearance of the pulse, the pressure of the cuff is raised another 30 or 40 mm Hg, and auscultation is performed while the cuff is slowly deflated. Blood pressure must be recorded in the opposite upper extremity even if the initial reading is normal. A unilateral atherosclerotic lesion in the right subclavian artery, for example, may result in a low right arm pressure; only when the blood pressure is measured in the left arm will the diagnosis of hypertension become evident. Another problem more frequently encountered in elderly patients is difficulty in determining the diastolic pressure. This results from the failure to hear the disappearance of the Korotkov sounds (phase V, which more accurately reflects diastolic pressure). In such cases, phase IV, the point of muffling, must be used. Any changes in blood pressure related to posture (orthostatic change) should be recorded.

E. Initiating Treatment: General Consideration

There continues to be considerable debate regarding when and how to treat hypertension in the elderly. Epidemiologically, elevated systolic and/or diastolic blood pressure predisposes to an increased incidence of cerebrovascular accidents, arterial aneurysms, renal insufficiency, multi-infarct dementia, and cardiac disease. Although there is conclusive evidence that normalization of blood pressure reduces these hypertension-related complications, little information is currently available regarding the benefits of lowering blood pressure in the elderly. In fact, only three hypertension studies have included elderly patients in their study populations. A double-blind, placebo-controlled study conducted by the Veterans Administration reported a reduction with treatment in cardiovascular complications in patients whose diastolic pressure on entry was 105 mm Hg or higher. Unfortunately, no subjects older than 69 were included. Also studying subjects to age 69, the Hypertension Detection and Follow-up Program compared treatment in specialized centers using protocol (stepped-care) treatment given by customary community resources. Although study design prevented a definitive conclusion, data suggested that lowering diastolic blood pressure to below 90 mm Hg reduced both mortality and the incidence of stroke.

The Australian Therapeutic Trial in Mild Hypertension randomized patients between the ages of 30 and 69 having diastolic pressures of 95 to 109 mm Hg into placebo and treatment groups. Treatment significantly reduced mortality. Although this was primarily related to a lower incidence of cardiovascular deaths, fewer cerebrovascular events were also noted in the treatment group.

Normalizing diastolic blood pressure appears to be effective in reducing associated mortality and morbidity at least in patients up to 69 years of age. Exactly what constitutes diastolic hypertension, however, remains controversial. Most recommend treatment when the diastolic pressure exceeds 100 mm Hg. Although there is little information regarding benefits of reducing diastolic pressure in those over age 70, most clinicians advocate control of blood pressure regardless of age.

An even more controversial problem is when to treat an isolated elevation in systolic blood pressure. Using the World Health Organization's criteria (systolic blood pressure >160 mm Hg and/or diastolic blood pressures >95 mm Hg), 40% of persons over age 65 are hypertensive; two-thirds of this group have an elevation of systolic pressure alone. Although one study reported the incidence of cardiovascular complications to relate independently to systolic as well as diastolic pressure, additional studies are being conducted to define better the consequences of isolated elevated systolic pressures. At the current time, most clinicians advocate treating isolated elevations of systolic blood pressure (>160 mm Hg) regardless of age.

F. Pharmacological Treatment

Many distinct classes of medications exist for the treatment of hypertension. Although all have the potential for side effects when given to patients of any age, the elderly are particularly prone to adverse reactions, especially those related to dosage. Changes in drug metabolism and clearance result from both the aging process and diseases that are more common at this time of life. Starting treatment with a low dose and slowly titrating upward until the desired effect is reached will avoid many complications. At times, however, side effects of a given medication must be carefully reviewed and weighed against potential benefits. (See table on pp. 46–47.)

It is important to remember that elderly hypertensive patients may require higher systolic pressures to maintain adequate blood flow through their less compliant vessels. Lowering blood pressure below a critical level (which may in a given person be well above 140 mm Hg) may result in inadequate cerebral perfusion. Carotid barorecepters, which are ordinarily able to compensate for this fall in blood pressure by inducing peripheral vasoconstriction and a faster heart rate, may be less sensitive in the elderly. For this reason, hypotension, especially related to positional change, is a common side effect of all classes of antihypertensive medications.

Older patients are uniquely less tolerant to many medications. Reduced plasma volume, perhaps related to an impairment in thirst perception, may predispose the elderly person to dehydration. Volume depletion from diuretic administration has an additive effect and can cause hypotension, which may or may not be orthostatic. Hypokalemia is another side effect of diuretics that may have more serious consequences in those elderly with preexisting cardiac disease. Careful monitoring of potassium levels and appropriate oral supplementation when necessary are essential.

β-Adrenergic blockers may also be problematic in the elderly

Antihypertensive Medications

Classification	Selected examples	Recommended initial dosage	Dose range (mg/day)	Frequency of administration/day	Side effects
Diuretics	Hydrochlorothiazide (Hydrodiuril and others)	25 mg q.d.	25–50	1–2	Volume depletion, hypokalemia, hyponatremia, hyperglycemia, hyperuricemia, hypercalcemia, anemia, photosensitivity, impotence
	Chlorthalidone (Hygroton® and others)	25 mg q.d.	25–50	1	
β blockers	Propranolol (Inderal®)	10 mg b.i.d.	20–480	2–3	Cardiac failure, bronchospasm, bradycardia, confusion, depression, impotence
	Metoprolol (Lopressor®)	50 mg q.d.	50–450	1–2	
	Nadolol (Corgard®)	20 mg q.d.	20–320	1	
Centrally acting agents	Clonidine (Catapres®)	0.1 mg b.i.d.	0.2–0.8	2	Dry mouth, sedation, gastrointestinal dysfunction, gynecomastia, impotence, abnormal liver function tests, anemia, lupuslike syndrome
	Methyldopa (Aldomet®)	125 mg b.i.d.	250–2000	2–3	
	Guanabenz (Wytensin®)	4 mg b.i.d.	8–32	2	

Arteriolar dilators	Hydralazine (Apresoline® and others)	10 mg q.d.	40–300	4	Tachycardia, peripheral neuritis, nasal congestion, lacrimation, blood dyscrasias, lupus (hydralazine), hirsutism (minoxidil)
	Minoxidil (Loniten®)	5 mg q.d.	5–100	1	
α blockers	Prazosin (Minipress®)	1 mg b.i.d.	2–20	2	Syncope, gastrointestinal dysfunction, vertigo, depression, impotence, dry mouth, nasal congestion
Converting enzyme inhibitor	Captopril (Capoten®)	6.25 mg b.i.d.	18.75–150	2–3	Proteinuria, neutropenia, rash, fever, dysgeusia, abnormal liver function tests
Calcium channel blockers	Nifedipine (Procardia®)	10 mg t.i.d.	30–120	3–4	Peripheral edema, cardiac failure, bradycardia, abnormal liver function tests
	Diltiazem (Cardizem®)	30 mg t.i.d.	120–240	3–4	
Peripheral sympatholytics	Guanethidine (Ismelin®)	10 mg q.d.	10–300	1	Cardiac failure, bradycardia, dizziness, diarrhea, inhibition of ejaculation, dry mouth, blurred vision, urinary incontinence, postural hypotension

with major risks of congestive heart failure, bradycardia, and/or heart block. In addition, they more commonly result in adverse effects on the central nervous system, causing fatigue, depression, and even hallucinations. A higher incidence of chronic obstructive lung disease in the elderly increases the risk of bronchoconstriction with these agents. Peripheral vascular insufficiency may also develop or become more symptomatic because of unopposed α stimulation. β_1-Selective agents such as metroprolol and combined β- and α-receptor blockers such as labetalol have not proved superior to nonselective agents (propranolol, nadolol, and others) in controlling blood pressure. Specific side effects must be considered in all cases.

Although centrally acting sympatholytic drugs are commonly used in the elderly, sedation and depression may limit their usefulness. Methyldopa has also been associated with sinus bradycardia, sinus pauses, 2 : 1 atrioventricular block, and carotid sinus hypersensitivity.

Hydralazine, a direct vasodilator, has unfortunately been associated with tachycardia relating at least in part to a chronotropic effect on the sinus node. Although this may potentially worsen ischemia in patients with coronary artery disease, the effect is thought to be minimized in the elderly because of baroreceptor insensitivity as described previously. Hydralazine can be safely administered as long as the dosage is carefully titrated and monitored and has the benefit of having minimal central nervous system side effects.

Prazosin, an antihypertensive with α-blocking properties, has not been tried extensively in the elderly. Reports of severe orthostatic hypotension in younger subjects, however, suggest that it be used only with extreme caution. Similarly, there has not been a great deal published regarding the use of captopril in the elderly. Recent reports, however, suggest that captopril has a good benefit-to-risk ratio in the elderly hypertensive patient. This drug prevents conversion of angiotensin I to angiotensin II and currently is recommended as a second-line antihypertensive agent. There have been some recent reports of success in safely treating older hypertensive patients with calcium channel-blocking agents. Despite this, myocardial depressant and conduction-blocking properties may limit their overall usefulness. Although recent studies have suggested that these agents be used early in the treatment of hypertension in the elderly, their use still remains controversial. Guanethidine, a peripheral catecholamine-depleting agent, often causes severe

orthostatic hypotension and should rarely be utilized in treating
older patients.

G. Summary

Until more definitive studies have been published, it appears
reasonable to attempt to keep blood pressures below 160 mm Hg
systolic and 95 mm Hg diastolic. Patients with borderline elevations
should be treated conservatively by recommending reduction of
dietary salt intake, encouraging weight loss, and recommending a
modest exercise program compatible with the patient's physical
capabilities and life style. The basic principle of prescribing antihypertensive agents is to follow a sequential pattern, beginning with
a low dose of one drug, while checking the patient frequently for
untoward effects; medication dosage should be increased slowly.
Other medications should be added as needed in a similar manner.
This is referred to as the stepped-care approach, recommended by
the Joint National Committee on Detection, Evaluation and Treatment of High Blood Pressure.

The usual first-line drug for managing hypertension in the
elderly (step 1) is a diuretic. This single medication is often all that
is needed to treat mild systolic and/or diastolic elevations. One of
the centrally acting agents, captopril or a β blocker can be
combined with the diuretic as the second-line choice (step 2). If
necessary, hydralazine (step 3) may be sequentially added for more
severe cases. Some advocate using hydralazine as a step 2 agent in
the elderly. Refractory patients may benefit from prazosin,
guanethidine, a calcium channel blocker or minoxidil, a potent
direct vasodilator, although the increased side effects with these
agents must be carefully considered and monitored.

Combination drugs should not be prescribed initially; their use
must be limited to those cases in which there is an exact match to
the effective regimen. Attempts must be made, however, to keep
the number of drugs to a minimum; frequency of administration
must be compatible with the patient's life style and habits. Non-
compliance becomes a major problem as numbers of medications
increase and dosing intervals become intolerable.

Hypertension is common in the geriatric population. Although
the precise benefit of treatment is unknown, extrapolation from
studies in younger age groups suggests that high blood pressure be
treated regardless of age. Hypertension must be identified early and

treatment promptly initiated to reduce morbidity and mortality. A suggested therapeutic goal is a systolic pressure of less than 160 mm Hg and a diastolic pressure of less than 95 mm Hg. Benefits of specific therapy must be carefully weighed against undesirable side effects. Antihypertensive medications must be carefully chosen, titrated, and monitored.

SUGGESTED READING

Cohen A.: Valvular heart disease in adults. *Postgrad Med* 74(4):299–306, 1983.

Cohn J. N.: Treatment by modification of circulatory dynamics. *Hosp Pract* Aug:37–52, 1984.

Duthie E.H., Gambert S.R., Tresch D.: Evaluation of the systolic murmur in the elderly. *J Am Geriatr Soc* 2(11):498–502, 1981.

Duthie D. H., Keeland M. H.: Geriatric cardiology and blood pressure, in Gambert S. R. (ed): *Contemporary Medical Geriatrics* (vol 2). New York, Plenum Press, 1986, pp 1–49.

Gersh B. J., Kronmal R. A., Schaff H. V.: Comparison of coronary artery bypass surgery and medical therapy in patients 65 years of age or older. *N Engl J Med* 313(4):217–224, 1985.

Gerson M. C., Hurst J. M., Hertzberg V. F.: Cardiac prognosis in noncardiac geriatric surgery. *Ann Intern Med* 103(6):832–837, 1985.

Gerstenblith G., Weisfeldt M. L., Kakatta E. G.: Disorders of the heart, in Andres R, Bierman E. L., Hazard W. R. (eds): *Principles of Geriatric Medicine*. New York, McGraw-Hill, 1985, pp 515–526.

Grossman E., Rosenthal T.: Infective endocarditis in the geriatric patient, *Geriatr Med Today* 4(7):50–57, 1985.

Miller D. C.: When to suspect aortic dissection: What treatment? *Cardiovasc Med* 1:811–818, 1984.

Phibbs B., Friedman H. S., Graboys T. B.: Indications for pacing in the treatment of bradyarrhythmias. *JAMA* 252(10):1307–1311, 1984.

Rodstein M.: The diagnosis of angina in the aged. *Geriatr Med Today* 2(10):80–86, 1983.

Schatz I. J.: Orthostatic hypotension *Arch Intern Med.* 144:773–777, 1984.

Schmidt S. B., Katz R. J.: PVCs in the elderly: When to start worrying. *Geriatrics* 40(8):30–44, 1985.

Topol E. J., Traill T. A., Fortuin N. J.: Hypertensive hypertrophic cardiomyopathy of the elderly, *N Engl J Med* 312(5):277–282, 1985.

3

Pulmonary Disease in the Elderly

I. THE AGING LUNG

Normal aging is associated with changes in the volumes and mechanics of pulmonary function. Representing one of the most integrated of physiological processes, pulmonary function and reserve are affected by age and further compromised by environmental factors and age-prevalent illnesses.

Age-Related Changes in the Pulmonary System

Decrease in elasticity of lung tissue
Increase in rigidity of thoracic cage
Decrease in strength of respiratory muscles
Decrease in forced expiratory vital capacity (FVC)
Increase in residual volume (RV)
Decrease in all flow rates
No change in total lung capacity (TLC)

A. Gas Exchange

Changes in pulmonary gas exchange may be noted with normal aging. Although arterial blood pH and $Paco_2$ remain unchanged, there is a gradual decline in arterial Pao_2. A decrease in systemic oxygen transport may result from declines in both oxygen saturation of hemoglobin and cardiac output. Diffusing capacity of the lung diminishes with age, primarily from a loss of lung tissue. This results in less total surface area available for alveolar–capillary interface.

With age, there is a decline in both central (medullary) and peripheral respiratory drive in response to changes in $Paco_2$ (hypercapnia) and Pao_2 (hypoxia), respectively. There is also a relative decrease in the cough response to noxious stimulus. The result is a weakened pulmonary defense against excess mucus secretion and aspiration, which may contribute to persistence of respiratory infection.

Smoking, at any age, will further compromise pulmonary defense mechanisms by increasing airway resistance, decreasing ciliary and alveolar macrophage function, and accelerating the normal age-related decline in pulmonary function.

B. Aspects of Pulmonary Disease in the Elderly

1. Diseases may be confused with elements of the normal aging process
2. There is great variability in disease presentation
3. Medications used in treatment must be carefully monitored, as side effects frequently are problematic in elderly patients

II. ASTHMA

Asthma is a chronic disorder characterized by an increase in sensitivity of the trachea and bronchi to a variety of stimuli. Reversible obstruction to air flow results from both smooth muscle contraction and mucosal edema of the airways. Acute attacks include **wheezing, cough,** and **dyspnea** of variable intensity and duration. Although attacks may frequently resolve spontaneously, acute medical intervention may be required.

Although most cases of asthma are diagnosed during youth, many elderly individuals have life-long disease and still require management. Presentation of illness in both previously diagnosed and undiagnosed older asthmatics may be atypical; often the only complaint is cough. A history of recent exposure to bronchial irritants may aid in the diagnosis.

Factors Capable of Precipitating an Acute Asthmatic Attack

1. Viral infection
2. Smoke
3. Polluted air
4. Cold air
5. Exercise
6. Use of aspirin or other nonsteroidal antiinflammatory agents

Since many other problems may result in wheezing, cough, and/or dyspnea, a thorough evaluation is essential. Differential diagnosis should include infection, congestive heart failure, pneumothorax, and pulmonary embolus.

Pulmonary function tests (spirometry or peak-flow meter) help assess both the severity of the acute episode and its response to therapy. An ECG may be necessary to rule out cardiac disease. Arterial blood gas measurements are important in the patient who is refractory to therapy or in whom there is fear of hypoxemia or CO_2 retention. Although a chest X ray is rarely necessary to assess asthma *per se*, it may help identify potentially causative factors, including infiltrates, congestive heart failure, or pneumothorax.

A. Therapy of Asthma

The therapeutic goal is to reverse bronchospasm and maintain airway patency and adequate ventilation. Although any identifiable extrinsic stimulus should immediately be removed from the environment, rarely can one be found.

Theophylline preparations and β-receptor agonists are frequently used medications for treating the elderly asthmatic. Although subcutaneous administration of epinephrine is a traditional and reliable form of therapy in the younger patient, problematic cardiovascular effects may limit its usefulness in the elderly. In choosing a β agonist, preference should be given to those having more β_2-agonist activity. β_1 agonists (e.g., epinephrine) have greater cardiac stimulatory effect; β_2 agonists (e.g., terbutaline) relax bronchial smooth muscle as well as blood vessels with less direct effect on the cardiovascular system. Unfortunately, even the β_2 agonists can cause significant side effects because of vasodilatation with resultant reflex tachycardia.

Medications (e.g., isoetharine, metaproterenol, and albuterol) are available for inhalation. These agents come in varying degrees of duration and β_2 selectivity. Since many elderly patients concomitantly use β-blocking medications for either heart disease, hypertension, or glaucoma, these drugs must be considered a potential cause of the asthmatic attack. If appropriate, measures should be taken to withdraw the β blocker slowly while introducing an alternative form of therapy.

Theophylline preparations may be useful for both acute management as well as maintenance therapy. These medications inhibit xanthine oxidase, increasing cAMP production with resultant bronchodilation. Careful attention to dosage is essential when using this medication in the elderly. A lower dosage is usually required in the elderly or those with liver disease or congestive heart failure. Concomitant cimetidine or erythromycin use lowers dose requirements; phenytoin and tobacco have been reported to raise dose requirements. Blood theophylline levels are useful to monitor therapeutic dosage. Hypoxemia may decrease clearance of theophylline.

Occasionally, systemic steroids may be required to control the acute asthmatic condition. In all cases, doses should be tapered as soon as possible to avoid numerous side effects including hypertension, gastric ulceration, central nervous system toxicity, decreased bone mineral content, and worsening of glucose tolerance. Inhaled steroids, although also capable of systemic side effects, may be better tolerated in the elderly patient.

B. Medications Used in Treating Asthma: Dosage and Characteristics

1. Bronchodilators

Parenteral
1. Epinephrine
   0.2–0.3 ml (1 : 1000 dilution) subcutaneous; rapid onset of action
2. Terbutaline (Bricanyl®)
   0.25 ml subcutaneous; relative delay of onset of action (30–60 min)

(continued)

1. Bronchodilators (Continued)

Inhalation

1. Isoetharine (Bronkosol®, Bronkometer®)
   0.25–0.5 ml in 2.5 ml normal saline via hand-held
   nebulizer (Bronkosol®); can be given every 2 hr as
   needed
   Also available in metered dose inhaler (Bronkometer®),
   two inhalations q4h
2. Metaproterenol (Alupent®, Metaprel®)
   0.2–0.3 cc in 2.5 cc normal saline via nebulizer; can be
   given every 4 hr as needed; a metered dose inhalation
   (two inhalations every 4–6 hr) is also available and
   useful for the outpatient
3. Albuterol (Proventil®, Ventolin®)
   Two inhalations as often as every 4 hr by metered
   inhaler (also portable); tends to be slightly longer
   acting than metaproterenol
4. Bitolterol (Tornalate®)
   Two or three inhalations every 6–8 hr by metered
   inhaler

2. Theophylline

Aminophylline (85% theophylline) is the usual parenteral form
used. A loading dose of 5 mg/kg body weight should be given i.v.
over 30 min or longer in severe cases, followed by a maintenance
dose usually between 0.2 and 0.5 mg/kg body weight per hour by
i.v. infusion. If the patient is already on oral theophylline, the
loading dose must be reduced or omitted.

Numerous oral theophylline products are available for use on a
one-, two-, three-, or four-times-per-day dosing schedule. Blood
levels help insure an optimal dosing regimen. In certain cases, peak
and trough levels may be beneficial to avoid toxicity at peak levels.
In addition, many elderly may be successfully tapered off of oral
theophylline preparations within weeks of the acute episode and
monitored either on nothing or p.r.n. inhalant therapy. This is
particularly true in the elderly person with wheezing secondary to
COPD.

3. Corticosteroids

An initial dose of 250–1000 mg i.v. hydrocortisone in the
acute setting should be followed by a maintenance dose of 100–300

mg i.v. every 4–6 hr. Patients may later be switched to oral forms of therapy; the dose should be tapered after the acute illness to the lowest amount still allowing good air flow. If possible, all therapy should be discontinued. If maintenance therapy is necessary, inhaled steroids (beclomethasone: Vanceril®, Beclovent®) via a metered dose inhaler can be an alternative to oral systemic steroids. Even in this form, long-term steroid treatment may lead to adrenal suppression.

III. CHRONIC OBSTRUCTIVE PULMONARY DISEASE

Chronic obstructive pulmonary disease (COPD) refers to elements of both chronic bronchitis and emphysema. In the elderly it is rare to see each as a distinct entity. The diagnosis of chronic bronchitis may be made after a productive cough has been present for at least 3 months in two successive years. The diagnosis of emphysema is based on pathological changes in lung parenchyma, with loss of lung tissue and an increase in residual lung volume.

Chronic obstructive pulmonary disease is associated with varying degrees of excess mucous production, mucus gland hypertrophy, mucus plugs, destruction of alveolar walls, and an overall reduction in expiratory airflow as measured by FEV_1. This latter change is more severe than would be predicted for age. Although the changes noted above occur over a period of many years, clinical disease is usually not problematic until the sixth or seventh decade of life. A history of ciagarette smoking is often obtained.

Early Signs of COPD

1. Cough
2. Wheezing
3. Recurrent respiratory infection
4. Dyspnea

Physical findings vary greatly. Most elderly patients with COPD will demonstrate decreased breath sounds, scattered wheezing and rhonchi, overinflation of the lungs, low diaphragms, and use of accessory respiratory muscles to varying degrees.

Therapy of COPD

1. Bronchodilator therapy may be useful if pulmonary function testing reveals a reversible component to airflow obstruction
2. Cigarette smoking should be eliminated or decreased as much as possible; even passive exposure to cigarette smoke may be problematic
3. Pulmonary toilet, when feasible, including postural drainage and cupping, may help eliminate secretions
4. Antibiotic therapy should be given for any sign of a significant change in sputum implying infection. Ampicillin (250 or 500 mg q6h for 10 days), tetracyline (250 mg q6h for 10 days), or trimethoprim plus sulfamethoxazole (Bactrim®, Septra® regular or DS, one dose q12h for 10 days) have proven useful when administered in this manner. Although some advocate culturing the sputum each time a change is noted, because of the numerous organisms infecting patients with COPD, many advocate empirical antibiotic therapy; cultures are usually reserved for refractory cases only
5. Diuretic therapy with or without digoxin should be reserved for use only in cases in which left ventricular function is diminished
6. Preventive therapy should include pneumococcal vaccine (Pneumovax®) as well as influenza vaccine. Current data suggest pneumonococcal vaccine lasts for at least 10 years if not for life; the influenza vaccine should be administered on a yearly basis each autumn
7. Cough suppressants should be avoided; this also includes medications with cough suppressant properties such as narcotics, certain tranquilizers (especially the phenothiazines), and hypnotics
8. Hydration should be encouraged to facilitate pulmonary toilet

IV. PULMONARY EMBOLISM

Pulmonary embolism is a frequent cause of death in the elderly. Autopsy findings exceed clinically reported cases, attesting to the difficulty in diagnosing this disorder. Since therapy exists for patients who survive the initial embolization, rapid diagnosis is essential.

Most pulmonary emboli originate in the deep veins of the pelvis and lower extremities, with venous stasis a causative factor.

Predisposing Factors

1. Prolonged immobilization (bed rest)
2. Varicose veins
3. Cancer
4. Estrogen use
5. Pelvic surgery
6. Congestive heart failure

There is often no prior history of leg pain or swelling. The classic symptoms of sudden onset pleuritic chest pain, dyspnea, and hemoptysis may be confused with other disorders in the older patient or be missed completely in the patient with prior cognitive dysfunction and/or dilirium. When these symptoms occur in the presence of a normal chest X ray, there should be a high degree of suspicion. Other clinical findings often seen include tachycardia, pleural rub, an increased pulmonic heart sound, rales, and fever.

Although specific changes may be noted on the ECG, they are frequently absent. The ECG is more useful in excluding myocardial infarction or ischemia.

A low Pao_2 may help make a diagnosis; however, age itself can lead to a lower Pao_2. In addition, pulmonary embolism may occur in the absence of an abnormal Pao_2.

The ventilation–perfusion lung scan helps the clinician diagnose a pulmonary embolus. Findings are reported as being either high, low, or of indeterminate probability. The patient with a clearly abnormal scan (unmatched perfusion–ventilation defects), clinical findings, and no other conflicting pulmonary or cardiac disease presents little diagnostic problem. In such a case, anticoagulant therapy is warranted. If diagnosis remains uncertain, as may occur in cases with concomitant pulmonary disease, a pulmonary angiogram may be required to confirm the suspicion. Since this test has a certain degree of morbidity, especially in the frail

elderly patient with chronic renal insufficiency and/or congestive
heart failure, the risk–benefit ratio must be carefully considered.

A. Treatment

Once a pulmonary embolus is diagnosed, intravenous heparin
should be given to prevent new clot formation. Therapy is initiated
with a bolus of 5000–10,000 units heparin to load and then
continued via i.v. infusion of a solution of ≤20,000 units/liter i.v.
fluid at rate of 1000 units/hr. In the absence of i.v. access or by
option, pulmonary embolism may be treated with heparin by
subcutaneous injection (5000 units q4h; 10,000 units q8h; or 20,000
units q12h). Therapeutic dose is obtained when partial thromboplastin time (PTT) is 1.5–2.5 times control. The PTT should be
followed closely and heparin dosage adjusted accordingly to insure
proper levels of anticoagulation. Patients should be placed on oral
anticoagulant therapy (coumadin) as soon as possible with the
prothrombin time (PT) maintained between 1.5 and 2 times control.
Therapy should be continued from 2 to 4 months with careful
attention given to monitoring for side effects. There does not appear
to be an increased risk of bleeding while on anticoagulant therapy
because of age alone. Risk factors, however, do include a number
of age-prevalent findings: renal and/or hepatic disease, recent
surgery, recent bleeding of gastrointestinal tract, polyps or ade-
nomas of colon, concomitant use of antiplatelet medications such as
aspirin, dipyridamole (Persantine®), or other nonsteroidal antiinflammatory drugs.

Several thrombolytic agents (streptokinase, urokinase) are now
available for use in the treatment of massive pulmonary emboli.
The risk–benefit ratio of these agents must be carefully considered.
Although they have been shown to improve the speed of clot
resolution, it is not clear whether they improve morbidity or
mortality in pulmonary embolism.

B. Prevention

In the bedridden patient, low-dose subcutaneous heparin (5000
units s.c. q8–12h) has been used prophylactically. This has been
shown to be a safe and potentially beneficial method of preventing
thrombotic events.

V. INTERSTITIAL LUNG DISEASE

Interstitial lung diseases include a group of over 130 diseases that are progressive and often fatal. The basic pathology involves alveolar inflammation and destruction with development of interstitial fibrosis. In some cases the cause is known (anthracosilicosis, asbestosis, farmer's lung, various drugs such as methotrexate). Onset of symptoms usually occurs between the ages of 30 to 50 years. Dyspnea on exertion, fatigue, malaise, and frequently a nonproductive cough are common. Sometimes there is a complaint of fever, night sweats, joint pain, and general flulike symptoms. Although initial auscultation may be normal, end-inspiratory rales may be heard as the fibrosis progresses. Far advanced disease may include clubbing or cor pulmonale. There are no specific lab findings. Pulmonary function testing demonstrates restrictive disease with a decrease noted in lung volumes, flow rates, and diffusing capacity.

Corticosteroid therapy is usually the only treatment available. In the elderly, only one-third of patients respond to this form of treatment.

VI. TUBERCULOSIS

Tuberculosis, though not as rampant in the western world as in the past, is still ever present. Recent reports confirm the significant persistence of this disease in the elderly, with an even greater prevalence in nursing home populations. In the elderly, disease is thought to be either recrudescence of infection acquired very early in life or repeat primary infection.

Group*	Active cases per 100,000
Overall	17
40–60 years old	23
>65 years living at home	60
>65 years in nursing home	234

*From Stead W. W., Lofgren J. P., Warren E., et al.: Tuberculosis as an endemic and nosocomial infection among the elderly in nursing homes. N Engl J Med 312:1483–1487, 1985.

Several authors have suggested that the increased incidence of active TB in the elderly at home represents a drop in immune system function. The even greater increase in nursing home populations represents the added effect of nosocomial infection. Other possible causes include malnutrition, glucose intolerance, neoplasia, and alcoholism. Although the classic presentation includes symptoms of night sweats, fever, loss of weight and appetite, and possibly hemoptysis, an atypical disease presentation is not uncommon. The chest radiograph often reveals infiltrates in the upper lobes, with possible cavitation. Presentation may even be in the form of miliary disease.

Diagnosis of TB in older patients requires a high degree of clinical suspicion. It must be seriously considered in elderly patients with unexplained cough or low-grade fever. Some may have a positive Mantoux skin test exceeding 9 mm in induration. Sputum cultures for TB must be done in all suspected cases, and necessary precautions taken to avoid spread in the community.

Current therapy for tuberculosis includes isoniazid, 300 mg/day, and rifampin, 600 mg/day, orally for 9 months. After 1 month, a change to a biweekly regimen may improve compliance (95% cure rate reported by Drs. Stead and Dutt in the volume *Principles of Geriatric Medicine*, Andres, R. (ed.), McGraw-Hill, New York, 1985). Unfortunately, although effective, both drugs have known hepatic toxicity and require surveillance of liver function. Recent studies have shown little benefit with the addition of ethambutol or streptomycin to this regimen except when faced with an INH-resistant strain.

Side Effects of Common Antituberculous Drugs

Drug	Potential side effect
Isoniazid	Peripheral neuritis; hepatotoxicity
Rifampin	Hepatitis; nausea, vomiting; flulike syndrome
Streptomycin	Vestibular nerve toxicity; nephrotoxicity
Pyrazinamide Ethambutol	Hepatotoxicity; optic neuritis
para-Aminosalicylic acid	Gastrointestinal; hepatotoxicity

SUGGESTED READING

Bardana C. J., Andrasch R. H.: Reviewing brochial asthma and its phar-
    macotherapy. *Geriatrics* 38:73–95, 1983.
Committee on Immunization: *Guide For Adult Immunization*. Philadelphia, American
    College of Physicians, 1985, pp 9–13.
Fulmer J. D., Snider G. L.: American College of Chest Physicians (ACCP).
    National Heart, Lung, and Blood Institute (NHLBI). Conference on Oxygen
    Therapy. *Arch Intern Med* 144:1645–1655, 1984.
Krumpe P. E., Knudson R. J., Parsons G., et al: The aging respiratory system, in
    Geokas M. C. (guest ed.): *Clinics in Geriatric Medicine*, (vol 1, no. 1).
    Philadelphia, W B Saunders, 1985, pp 145–175.
Mahler D. A.: Pulmonary aspects of aging, in Gambert S. R. (ed): *Contemporary
    Geriatric Medicine* (vol 1). New York, Plenum Press, 1983, pp 45–85.
Nagami P. H., Yoshikawa T. T.: Tuberculosis in the geriatric patient. *J Am Geriatr
    Soc* 3:356–363, 1983.
Rudd A.: Tuberculosis in a geriatric unit. *J Am Geriatr Soc* 33(8):566–569, 1985.
Stein P. D., Willis P. W.: Diagnosis, prophylaxis, and treatment of acute pulmonary
    embolism. *Arch Intern Med* 143:991–994, 1983.

4

Diseases of the Gastrointestinal Tract

I. INTRODUCTION

Diseases of the gastrointestinal tract (GIT) are common in the elderly and are responsible for approximately 30% of visits that older patients make to the physician. Normal aging is also associated with an alteration in function in almost all aspects of the gastrointestinal tract. Altered motility of the esophagus, atrophy of gastric mucosa with resultant reduction in acid secretion, constipation, and diverticular disease are just some of the changes that are commonly seen in the elderly relating to both "normal" aging and age-prevalent illness. Some of these changes have been attributed to an altered life style, i.e., a diet deficient in calories, protein, fiber, and fluid intake, and retirement, loneliness, loss of self-esteem, and anxiety.

Effect of Age and Illness on the Gastrointestinal System	
Location	Changes/diseases
Mouth	Loss of taste buds
	Decreased salivary secretion
	Loss of dentition
Esophagus	Altered motility
	Reflux esophagitis
	Carcinoma

(continued)

Effect of Age and Illness on the Gastrointestinal System
(Continued)

Location	Changes/diseases
Stomach	Reduction in acid secretion
	Delayed gastric emptying
	Hiatus hernia
	Peptic ulcer
	Polyps
	Carcinoma
Small intestine	Mucosal atrophy
	Impaired absorption of calcium
	Lactase deficiency
	Diverticulosis
Large intestine	Poor peristalsis
	Constipation
	Diverticulosis
	Ischemic colitis
	Carcinoma
	Rectal prolapse

II. HIATUS HERNIA AND REFLUX ESOPHAGITIS

It has been estimated that between 50 and 70% of all people over the age of 70 have a hiatus hernia. In addition, of those with a demonstrable hiatus hernia, almost half will occasionally have some degree of esophageal reflux with symptoms. Changes in life style including overeating, smoking, and obesity have been significantly associated with an increased incidence of hiatus hernia and reflux esophagitis. Weakness of the lower esophageal sphincter is thought to be the primary factor responsible for gastroesophageal reflux.

A. Clinical Features

Symptoms of hiatus hernia and esophageal reflux may include chest pain, heartburn, and difficulty in swallowing. It is important to remember that although reflux is commonly associated with hiatus hernia, both reflux and hiatus hernia may occur independently. Some of the special features of hiatus hernia and reflux disease in the elderly are listed.

Special Features of Hiatus Hernia and Reflux Esophagitis in the Elderly

1. Symptoms may escape recognition; anemia caused by blood loss and/or dysphagia may be the only presenting problems
2. Aspiration of gastric contents may result in coughing, wheezing, morning hoarseness, or even pulmonary infection
3. A chest X ray is often indicated when aspiration is suspected
4. If an esophageal stricture and/or dysphagia is noted, further evaluation and histological confirmation are essential to rule out malignant disease

"Heartburn" is a specific symptom related to reflux that may be aggravated by fatty meals, change in position such as lying down or bending to tie one's shoelaces, a heavy meal, or tight undergarments. This may result in an increased lower abdominal pressure exceeding lower esophageal sphincter limits. It is important to remember that occasional reflux is a normal phenomenon; reflux should be considered significant only if the patient complains of recurrent symptoms. Although hiatus hernia may present as dysphagia, numerous other causes must be excluded. Dysphagia often presents diagnostic difficulties requiring radiological delineation. Common causes of dysphagia in old age are listed.

Common Causes of Dysphagia in Old Age

Condition	Comments
1. Painful conditions of the oropharynx	Tooth abscess
	Glossitis/pharyngitis
	Leukoplakia
	Carcinoma of tongue
2. Edentulous patient	Altered swallowing ability
3. Decreased salivation	Sjögren's syndrome; medication-related changes
4. Esophagus/stomach	Esophagitis
	Stricture
	Carcinoma
	Achalasia
	Plummer–Vinson syndrome
	Systemic sclerosis
	Abnormal esophageal motility

(continued)

Common Causes of Dysphagia in Old Age (*Continued*)

Condition	Comments
5. Neuromuscular	Bulbar/pseudobulbar palsy
	Myasthenia gravis
	Parkinsonism
	Dystrophia myotonia
6. Psychogenic	Anxiety neurosis
	Depression

B. Differential Diagnosis

Pain associated with reflux esophagitis must be clearly distinguished from other causes, particularly related to ischemic heart disease. In both cases the presenting symptom may be a sharp or crushing substernal pain. Emotional upset may precipitate the pain in both cases. The following table summarizes salient features to help distinguish between these two types of pain. Although these two entities must be carefully distinguished, many elderly patients have both conditions.

Differential Diagnosis of Chest Pain

Esophagitis	Ischemic heart disease
Substernal	Substernal
May be precipitated by emotion	May be precipitated by emotion
Worse on taking tea/coffee	No effect of tea/coffee in most cases
Alleviated by antacids	No effect of antacids
May improve with change in posture	Change in position has no effect
Usually not exacerbated by exertion	Usually worsened by exertion
Acid perfusion test with ECG monitoring usually diagnostic	Acid perfusion test with ECG monitoring usually diagnostic

---

C. Management of Hiatus Hernia and Reflux Esophagitis

Reduction in body weight, cessation of smoking, and avoidance of large fatty meals usually help reduce symptoms related to the hiatus hernia with reflux. Smoking causes relaxation of the lower esophageal sphincter, leading to reflux. Alcohol increases gastric acid secretion, also favoring gastroesophageal reflux. Elderly persons who are accustomed to drinking an alcoholic beverage in the evening, for example, may need to change their timing; symptoms of reflux are particularly troublesome at night. Other measures that may help in treating this disease are listed.

Nonpharmacological Treatment of Hiatus Hernia with Reflux Esophagitis

1. Correction of obesity, i.e., weight loss to within 15% of average weight for age
2. Avoid late evening meals
3. Avoid large meals, especially with high fat content
4. Avoid excessively hot drinks
5. Avoid alcohol/caffeine
6. Stop smoking
7. Head of bed should be raised on 4- to 6-inch blocks
8. Avoid multiple pillows or "sitting up" position during night
9. Avoid wearing constricting garments around the abdomen
10. Avoid bending over

Antacids are the most commonly used treatment for hiatus hernia and reflux esophagitis. In the older patient, it is important to consider side effects when choosing an appropriate antacid. Calcium-containing antacids may exacerbate constipation, a problem not uncommon in the elderly. Magnesium-containing antacids may, on the other hand, cause diarrhea, increasing the risk of dehydration and electrolyte imbalance. Aluminum-containing antacids must be used with caution in the presence of poor renal function. Besides antacids, H_2 receptor blockers such as cimetidine (Tagamet®) and ranitidine (Zantac®) have been found to be helpful in the treatment of hiatus hernia and reflux esophagitis. The use of antacids and H_2 receptor antagonists is summarized under the pharmacological treatment of peptic ulcer disease in the elderly (Section III.A).

III. PEPTIC ULCER DISEASE IN THE ELDERLY

Peptic ulcer disease affecting both the gastric and duodenal mucosa is not uncommon in old age. Its presence, however, is probably underestimated by many physicians. The disease often presents atypically and nonspecifically.

Common Features of Peptic Ulcer Disease in the Elderly

1. Epigastric pain: poor localization of pain is common; pain may not be related to the intake of food
2. Nausea and vomiting
3. Loss of appetite and weight not infrequently noted; these symptoms are often confused with a diagnosis of neoplastic disease
4. Flatulence
5. Regurgitation in the mouth with a clear tasteless fluid may occur
6. Heartburn: this may be confused with cardiovascular disease or hiatus hernia with reflux
7. Anemia: this may result from chronic blood loss; occasionally the patient may present with an acute gastrointestinal hemorrhage or perforation even in the absence of a clear history of dyspepsia

---

Although the exact etiology of peptic ulcer disease in the elderly remains unclear, the incidence of gastric ulcer increases with age. A comparatively lower incidence of duodenal ulcers in the elderly may be related to decreased gastric acid secretion by the parietal cells of the stomach. History, clinical examination, radiological testing, and endoscopy are helpful in localizing the site of the ulcer.

Distinguishing Features of Gastric and Duodenal Ulcer

	Gastric ulcer	Duodenal ulcer
Incidence	Increases with age	Decreases with age
Epigastric pain	Pain usually appears within 1 hr of meals and lasts up to 2 hr	Pain reduced by food: maximal pain usually noted 2 to 3 hr after meals
Eating habits	Pain may intefere with eating, and skipped meals are common	Frequent meals common
Weight	Weight loss may be noted	Weight gain not infrequent
Acidity	High-acid state uncommon	Hyperacidity common

Elderly patients with suspected peptic ulcer disease should be evaluated if symptoms persist. Almost all elderly patients with gastric ulcer disease undergo endoscopy. In most cases biopsy and histological confirmation of the lesion are essential in order to rule out malignancy. This is particularly important because of the high incidence of gastrointestinal malignancy in the elderly, particularly in those with low gastric acid secretion. (See table on p. 71.)

A. Drug Therapy for Peptic Ulcer Disease

In recent years, H_2-receptor antagonists (cimetidine and ranitidine) have been widely used in the treatment and prevention of both gastric and duodenal ulcers. Although elderly persons have a higher incidence of frank achlorhydria, few bother to test gastric pH prior to initiating this form of therapy. Little is known about whether these agents are effective in the achlorhydric population: side effects must be carefully looked for. Although these agents are no more effective in lowering acid secretion than antacid therapy, compliance is somewhat better. (See table on p.70.)

B. Complications of Peptic Ulcer Disease

Complications of peptic ulcer disease in the elderly are not significantly different from those seen in younger adults. Serious complications include

Pharmacological Treatment of Peptic Ulcer Disease in the Elderly

Type of medication	Proposed mechanism of action	Dose/day	Potential side effects
Antacids: usually contain sodium bicarbonate, calcium carbonate, aluminum or magnesium salts	Neutralize gastric acid	1–2 tablets or 15–30 ml every 2–6 hr	Sodium overload with water retention; worsening of heart failure with sodium bicarbonate. Hypercalcemia, decreased renal function and milk-alkali syndrome with calcium carbonate. Aluminum salts may cause neurotoxicity, phosphorus depletion, and osteomalacia. Magnesium salts may cause diarrhea, hypotension, respiratory depression, and renal damage
Sucralfate (Carafate)	Antipepsin effect; acts by forming a protective barrier at the ulcer site	1 g four times a day	Constipation, nausea, diarrhea, gastric discomfort, dry mouth, rash, pruritis, dizziness, sleepiness, and vertigo
Cimetidine (Tagamet®)	H_2-receptor antagonist, i.e., blocks gastric acid secretion at the level of parietal cell; inhibits both daytime and nocturnal basal gastric acid secretion	400–1200 mg per day in divided doses	Reversible confusional states, e.g., mental confusion, agitation, psychosis, depression, anxiety, hallucinations; gynecomastia, impotence, transient diarrhea, dizziness, somnolence, rashes, dementia; drug interactions common
Ranitidine (Zantac®)	As above	150–300 mg in divided doses	Skin rash, malaise, dizziness, insomnia, vertigo; rarely, mental confusion, agitation, depression, hallucination, constipation, and diarrhea

Managing Suspected Peptic Ulcer Disease in an Elderly Patient

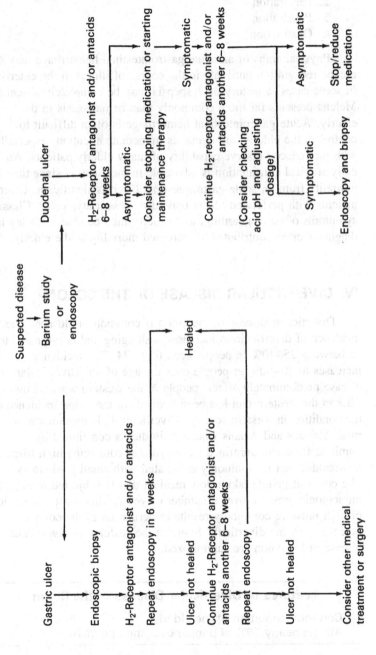

1. Hemorrhage
2. Perforation
3. Penetration
4. Obstruction

Physical signs of an acute gastrointestinal hemorrhage may escape recognition until late in the course of illness in the elderly. In some cases, a history of dyspepsia may be completely absent. Melena presents far more commonly than hematemesis in the elderly. Acute gastrointestinal hemorrhage is often difficult to control in the elderly and demands immediate attention, especially with the reduced reserve capability of many elderly patients. An early surgical consultation is advisable in most cases, since the mortality from an acute gastrointestinal bleed is significant. Elderly patients with perforated ulcers may also present atypically. Classic symptoms of pain, guarding, and rigidity may be absent. Delay in diagnosis often contributes to increased mortality in the elderly.

IV. DIVERTICULAR DISEASE OF THE COLON

Diverticular disease of the colon is common in old age. The incidence of diverticulosis increases with aging and is estimated to be between 25–40% in people aged 65 to 74. The incidence increases to 40–50% in people over the age of 75. Diverticular disease predominately affects people in the western world. Low fiber in the western diet has been blamed for the rising incidence of the condition in western society. Diverticulosis is uncommon in rural Africans and Asians whose main diet is constituted by unmilled flour and unrefined carbohydrates containing high fiber. A low-residue diet is commonly associated with small hard stools. The colon of people taking low residual diet is subjected to high intracolonic pressure over a number of years. This constant exertion of high pressure commonly results in herniation of the colonic mucosa, i.e., the diverticuli. Some of the features of diverticular disease of the colon are summarized.

Features of Diverticular Disease of the Colon

Common condition of the elderly
Affects nearly 50% of people over the age of 70

(continued)

Features of Diverticular Disease of the Colon (*Continued*)

More common in the females than males

Most diverticuli are diffusely distributed in sigmoid colon

As the number of diverticuli increases so do the chances
of symptoms and complications

Solitary diverticula may sometimes affect the cecum and
the ascending colon (if carcinoma cannot definitely be
excluded such patients may require surgery)

Clinical presentation of the diverticular disease includes various symptoms. Change in the bowel habits is perhaps the commonest symptom. Constipation, diarrhea, or constipation alternating with diarrhea may be seen. Since a similar change in bowel habit may result from carcinoma of colon as well, the physician should spare no efforts in the correct assessment and investigation of an elderly patient presenting with change in bowel habit. Symptoms of diverticular disease in old age are given.

Presentation of Diverticular Disease in Old Age

Abdominal discomfort/pain

Change in bowel habits; diarrhea/constipation; diarrhea
alternating with constipation

Flatulence and heartburn

Nausea and vomiting

Rectal bleeding

Weight loss

Palpable mass in left lower quadrant

Asymptomatic finding of diverticuli in barium enema carried out for some other gastrointestinal disorder

Several conditions have been found to be significantly associated with diverticular disease of the colon. Some of these conditions are listed.

Conditions Commonly Associated with Diverticular Disease

Hiatus hernia

Gall bladder disorder

(*continued*)

Conditions Commonly Associated with Diverticular Disease (Continued)

Coronary artery disease
Hemorrhoids
Varicose veins

A high-fiber diet (>30 g daily) can be useful in both the prevention and treatment of diverticular disease. Data suggest that the regular intake of bran and increased consumption of raw fruits and vegetables are important for people in their middle and late years.

Food with High Fiber Content

Food source	Dietary fiber/100 g
Cereals	
Bran	44.0
All-Bran™	29.9
Corn bran	16.1
Wheat Chex™	12.7
40% bran flakes	14.1
Raisin bran	10.8
Honey bran	11.1
Nuts	
Coconut (dried)	23.5
Peanuts	8.1
Almonds	14.3
Fresh fruits	
Blackberries	4.5
Cranberries	3.0
Raspberries	4.7
Fruits cooked/canned	
Prunes, stewed	8.1
Guava, canned	3.6
Vegetables	
Spinach	6.3
Broccoli	4.1
Peas	6.3
Corn	5.7
Carrots	3.1
Raw cabbage	3.4
Cooked leeks	3.9
Raw mushrooms	2.5

A. Treatment of Diverticular Disease

Steps necessary in the prevention and treatment of diverticular disease are listed.

1. High-fiber diet
2. Avoid drugs (narcotics, antacids containing aluminum or calcium, anticholinergic agents) that cause constipation
3. Encourage plenty of fluids
4. Avoid sedentary life style
5. Use stool softeners as required
6. Avoid cathartics and purgatives

B. Complications of Diverticular Disease

It is estimated that between 10% and 30% of patients with diverticular disease have one or more complications, the majority of which occur in late life. Infection and inflammation of the diverticula, i.e., diverticulitis, is often associated with the blockage of the diverticulum with feces. The majority of complications can be effectively treated medically, especially if the diagnosis is made early in the course of illness. Surgery may be necessary and can be life saving in certain situations. Various complications and treatment of diverticular disease are summarized.

Complications and Treatment of Diverticular Disease

Complication	Clinical features	Treatment

Acute diverticulitis	Abdominal pain
	Localized
	  tenderness
	Fever
	Leukocytosis
		Nasogastric suction
		Intravenous fluids
		Systemic antibiotics
		  directed towards
		  gram-negative or-
		  ganisms

Pericolic abscess	Abdominal pain
	Fever
	Leukocytosis
	Palpable mass, usu-
	  ally in the left
	  lower quadrant
		All the above steps
		  are necessary
		Surgery may be re-
		  quired

Perforation of the
  diverticulum
	Abdominal pain
	Guarding/rigidity
	Air under the di-
	  aphragm on ab-
	  dominal X ray
		Conservative man-
		  agement includes
		  treatment of
		  shock with fluids
		  and antibiotics
		Surgery may be re-
		  quired

(continued)

Complications and Treatment of Diverticular Disease (*Continued*)

Complication	Clinical features	Treatment
Intestinal obstruction	Crampy abdominal pain and distention Constipation Vomiting	Nasogastric suction Intravenous fluids Surgery may be required
Fistula (vesical or vaginal) formation	Recurrent urinary tract infections "Fecal" soiling	Surgery is required in all cases

---

V. INFLAMMATORY BOWEL DISEASE

Ulcerative colitis and Crohn's disease are both inflammatory diseases of the gastrointestinal tract of unknown etiology. Although both diseases may occur at any age, they predominantly affect younger adults. Several other forms of colitis that commonly affect the elderly include ischemic colitis, nonspecific proctitis, and drug-induced (especially broad-spectrum antibiotics) colitis.

A. Ulcerative Colitis

Various features of uncomplicated ulcerative colitis include:

1. Diarrhea with or without blood/mucous.
2. Lower abdominal pain.
3. Constitutional symptoms, e.g., anorexia, weight loss, anemia, lethargy, and tiredness.

As in some other conditions affecting the elderly, diagnostic difficulties may be encountered if ulcerative colitis first presents in old age; an atypical presentation is not uncommon. Diarrhea, blood loss, anorexia, and weight loss may sometimes be attributed to ischemic colitis, diverticular disease, or carcinoma of the large intestine. Although ulcerative colitis is less likely to affect the elderly, it must be considered in all elderly patients with prolonged and unexplained diarrhea.

Sigmoidoscopy is diagnostic in most cases. Barium enema, although almost always required to assess the extent of the disease, should not be performed during acute illness. Diagnostic features seen on sigmoidoscopy and radiology are listed.

Characteristics of Ulcerative Colitis

Sigmoidoscopic examination	Radiologic findings on barium
Red and inflamed mucosa	Narrowing and shortening of bowel
Contact bleeding	Poor distensibility of colon
Granular appearance	Edema of the wall
Edema	Diffused ulceration with "fuzzy" appearance
Abnormal vascular pattern	Loss of haustration
	Polyps

Medical treatment of unremitting ulcerative colitis includes the use of steroids (prednisone, 30–60 mg in divided doses) and sulfasalazine (Azulfidine, 2–4 g daily in divided doses). Although steroids are indicated for the treatment of an acute inflammation, sulfasalazine is reserved for maintenance treatment. In the older patient, an acute attack must be treated promptly with steroids given both parenterally and in an enema. Special attention must be given to fluid and electrolyte loss in the elderly. After control of the symptoms is achieved, the daily doses of steroids may gradually be reduced by 5 to 10 mg every week. Gradual tapering off of the steroids is successfully achieved in almost all cases. Careful monitoring of side effects of steroids is essential in the elderly: side effects may include sodium and water retention, potassium loss, muscle weakness, osteoporosis, and diabetes mellitus.

Although surgery is rarely required for ulcerative colitis, indications include:

1. Failure of medical therapy, i.e., if the patient continues having symptoms of anemia, weight loss with diarrhea, and blood loss.
2. Toxic megacolon.
3. Evidence of malignancy or premalignant change.
4. Prophylatic colectomy in longstanding colitis.

Complications of ulcerative colitis relate to the extent of colonic involvement and the frequency and severity of acute attacks. Potential complications are listed below.

Complications of Ulcerative Colitis in Older Patients

1. Nutritional deficiency
   a. Anemia
   b. Dehydration
   c. Hypoproteinemia
   d. Vitamin deficiency
2. Skin and mucous membrane
   a. Erythemia nodosum
   b. Erythema multiforme
   c. Leg ulcers
   d. Pyoderma gangrenosum
3. Local complications
   a. Hemorrhage
   b. Toxic megacolon
   c. Perforation
   d. Ischiorectal abscess
   e. Fissure formation
   f. Stricture
   g. Fistula
   h. Pseudopolyposis
   i. Carcinoma
4. Arthritis
5. Uveitis
6. Amyloidosis
7. Miscellaneous
   a. Complications of blood transfusion
   b. Complications of steroid therapy

B. Regional Ileitis (Crohn's Disease)

Although Crohn's disease presents most commonly during early adult life, another peak is noted during late life. In older patients, Crohn's disease more commonly affects the large bowel. Diarrhea is the most common early symptom; pain, particularly before defecation, low-grade pyrexia, anemia, weight loss, and arthritis are also common. Similarities between Crohn's disease and ulcerative colitis may create a diagnostic dilemma.

Crohn's Disease and Ulcerative Colitis: A Comparison

	Ulcerative colitis	Crohn's disease
Etiology	Not known	Not known
Pathology	Primarily mucosal involvement with polymorphonuclear, plasma cells and eosinophilic infiltration	Patchy transmural involvement Infiltration with lymphocytes, plasma cells, and macrophages is characteristic

(*continued*)

Crohn's Disease and Ulcerative Colitis: A Comparison (*Continued*)

	Ulcerative colitis	Crohn's disease
	Destruction of mucosal glands is common	Glandular presentation is common
	Fibrosis is uncommon	Significant fibrosis is common
Clinical features	Bloody diarrhea common	Bloody diarrhea rare
	Anemia, weight loss, fever, and features of malabsorption are less common	Anemia, weight loss, fever, and features of malabsorption are common
	Usually no abdominal mass is present	Abdominal mass is common
Barium studies	Continuous disease	Patchy segmental involvement
	Fine ulceration	Coarse lesions noted, giving a "cobblestone" appearance
	Rectum involved in most cases	Rectum spared in most cases
Treatment	Steroids	Steroids not usually indicated
	Sulfasalazine	Sulfasalazine
Complications	Rare	Fistula, stricture, and rectal abscess common

VI. GASTROINTESTINAL HEMORRHAGE

Acute gastrointestinal hemorrhage is not an uncommon problem in the elderly, leading to emergency admission to the hospital. Bleeding from peptic ulcer disease accounts for approximately 70% of these cases. The remaining 30% result from esophageal varices, gastritis, carcinoma of stomach, and the Mallory–Weiss syndrome (esophageal tears). Elderly patients may respond atypically to the blood loss; i.e., the compensatory rise in pulse rate that normally accompanies volume depletion during youth may not occur in the older patient. Furthermore, as many as 25% of cases present without any prior history of gastrointestinal distress. Melena occurs more commonly than hematemesis in the elderly.

Special Features of Gastrointestinal Hemorrhage in the Elderly

1. History of previous "gastrointestional distress" not always noted

(continued)

Special Features of Gastrointestinal Hemorrhage in the Elderly (*Continued*)

2. Pulse and blood pressure changes may be misleading
3. Melena is more common than hematemesis
4. Rebleeding within 48 hr of first bleed is more common
5. Early surgical referral is necessary in most cases
6. Mortality is significantly higher compared to young adults

General principles of the management of gastrointestinal hemorrhage should not be altered as a function of age. Since elderly patients tend to bleed for prolonged periods, however, surgical consultation should be sought at an early stage. Increased incidence of rebleeding and a relatively high mortality rate must be kept in mind. In general, supportive measures are essential, with hemodynamic monitoring, fluid and electrolyte balance, and avoidance of either volume depletion or overhydration. Because of decreased physiological reserve and problems in maintaining homeostasis, elderly patients with gastrointestinal bleeds deserve strict supervision and early treatment. Following stabilization, endoscopic and angiographic (if available) evaluation should be pursued. The use of intraarterial vasopressin infusion in the bleeding vessel may reduce the blood loss: 5–20 units of vasopressin, repeated every 4–6 hr, injected intraarterially may help reduce hemmorrhage by constricting splanchnic arterioles. Side effects of vasopressin therapy include nausea, intestinal colic, uterine cramps, defecation, and constriction of coronary arteries, capable of precipitating an anginal attack. Vasopressin, therefore, must be used cautiously and in low doses in the elderly.

VII. CONSTIPATION IN THE ELDERLY

Although most people believe that constipation is to be expected as one ages, considerable natural variation makes it extremely difficult to define what actually constitutes constipation. Most elderly complain of constipation because they believe a daily bowel movement is the only pattern consistent with good health. The fear of becoming constipated is probably an even greater problem than the actual condition itself; one-fifth of elderly persons claim to be constipated when questioned; one-half take laxatives on a regular basis.

"Constipation" may represent either a difficulty in defecating because of hard stools or less frequent but naturally formed bowel movements. In cases of severe constipation with fecal impaction, bowel movements may actually occur at regular intervals; diarrhea occurring around the fecal obstruction may also be noted.

When one is confronted with this common problem, a differential diagnosis must be considered in order that any serious underlying disorder not be missed.

Common Causes of Constipation in the Elderly

Psychological	Depression
	Dementia
Obstructive	Neoplasms
	Stricture
	Adhesions
	Megacolon
Metabolic	Myxedema
	Hypokalemia
	Hypercalcemia
	Dehydration
Painful anal lesions	Thrombotic/inflamed
	hemorrhoids
	Anal fissure
Iatrogenic	Anticholinergic agents
	Antacids containing alumi-
	num or calcium
	Narcotics
	Any medication with anti-
	cholinergic properties

Proper diet with adequate bulk and fluid intake helps in the prevention of constipation. In addition, laxatives may be used to treat constipation when natural methods fail.

Laxatives Used Orally for Treatment of Constipation

Name: Generic (brand)	Usual daily dose	Comments
Bulk-forming agents		
Bran	12–24 g daily in divided doses	May interfere with absorption of calcium, iron, and digoxin; bloating can result

---

(continued)

Laxatives Used Orally for Treatment of Constipation (*Continued*)

Name: Generic (brand)	Usual daily dose	Comments
Byllium Hydrophilic mucilloid (Metamucil®)	1 packet (7 g) 2–3 times a day	Adequate fluid intake essential; do not use in cases of fecal impaction or suspected intestinal obstruction
Stool softeners		
Docusale sodium (Colace®)	50–200 mg	Mild abdominal cramps and skin rashes may occur; excess use can lead to diarrhea
Liquid paraffin	10–30 ml	Can cause lipid pneumonitis secondary to oil aspiration
Stimulant laxatives		
Cascara sagrada	200–400 mg	Long-term use may cause hypokalemia, hypocalcemia, protein-losing enteropathy, diarrhea, and malabsorption; coincidental use of antacid may dissolve coating and cause abdominal cramps
Senna (Senokot®)	1–2 tabs (187–347 mg)	
Bisacodyl (Duccolax®)	5–15 mg	
Osmotic laxatives		
Milk of magnesia	30–60 ml	Not to be used in patients with renal impairment; excess use can affect CNS and neurological functioning
Lactulose (Chronulac®)	15–30 ml	Preferably given with milk or juice; acts both as an osmotic agent and as a stimulant of bowel motility; rare potential for hyperglycemia

Suppositories (glycerin and bisacodyl) cause mucosal irritation leading to propulsion of stools. Enemas (tap water, saline, or oil) help in the treatment of constipation by inducing reflex evacuations.

Fecal incontinence is perhaps the most common complication of constipation and fecal impaction. Many other potential complications exist.

Potential Complications of Constipation

Fecal incontinence with diarrhea Megacolon
Delirium Urinary retention/incontinence
Large bowel obstruction Laxative abuse

VIII. FECAL INCONTINENCE

Fecal incontinence is most frequently seen in institutionalized elderly populations, where it has been estimated to occur in as many as 15 to 30% of patients.

Common Causes of Fecal Incontinence

1. Fecal impaction   4. Dementia or delirium
2. Diarrhea due to any cause   5. Neurological impairment
3. Rectal prolapse

When incontinence is caused by fecal impaction, treatment of the underlying condition with a cathartic, enema, or manual removal is essential; steps to prevent recurrence must be initiated. Increasing fiber in the diet, exercise as tolerated, insuring adequate hydration, and use of fecal softening agents when required should be considered. Rectal prolapse may require surgical intervention.

A toileting program similar to that used in the patient with urinary incontinence may help eliminate this problem. Fecal incontinence is more commonly seen in the patient with urinary incontinence, and a combined toileting approach is often effective. Complications of fecal incontinence include social isolation, depression, skin breakdown, infection, and increased laundry costs.

IX. GALLBLADDER DISEASE

Patients over 65 years of age have a startling sixfold increase in mortality rate during surgery for acute cholecystitis. Increased prevalence of diabetes, chronic lung disease, and coronary artery disease in the elderly are thought to be contributing factors to this excessive mortality. Despite this, elective surgery in the elderly for chronic cholecystitis carries a mortality rate of only 2%. A greater choice of diagnostic interventions and improved antibiotics have reduced the number of unnecessary exploratory laparotomies; percutaneous or endoscopy-directed therapies also may eliminate the need for surgery.

Biliary stone formation increases with age and estrogen use. Women have a two to three times higher incidence of cholesterol gallstones than men. A more fulminant course has been seen in men than in women in acute cholecystitis and choledocholithiasis. Should the use of estrogen become widespread in the treatment of osteoporosis, an increased incidence of stone formation will most certainly result. Three factors cause increased cholesterol saturation of bile in the elderly:

1. Decreased synthesis of bile acids.
2. Reduced cholic acid pool size.
3. Increased hepatic secretion of cholesterol.

In a recent study, no significant difference between the sexes was noted when each of these factors was analyzed. This refutes earlier claims of a hormonal–endocrine etiology of gallstone formation.

Stages of Acute Cholecystitis

1. Cystic duct obstruction
2. Cholesterol saturation of bile leading to chemical irritation and inflammation of the gallbladder (wall thickening on ultrasound)
3. Copious secretion of mucous (distention of the gallbladder causing edema of the gallbladder wall)
4. Interference with lymphatic and venous drainage of the gallbladder causing edema of the gallbladder wall
5. Compromised arterial blood supply resulting in patchy necrosis and gangrene of the gallbladder wall (anechoicity on ultrasound)

A. Clinical Features of Cholecystitis/Cholelithiasis in the Elderly

Feature	Possible cause
1. Right upper quadrant or epigastric pain; colic	Biliary obstruction
2. Nausea and vomiting; distention of the gallbladder or common bile duct	Gallstone with or without cholecystitis
3. Fever with or without bacteremia	Infection and/or dehydration
4. Acute abdomen/peritonitis and/or shock versus palpable gallbladder with localizing peritoneal signs	Rupture of the viscus
5. Other: jaundice, pancreatitis	Ductal obstruction

B. Available Diagnostic Procedures

1. Oral cholecystography.
2. Hepatocholescintigraphy.

3. Ultrasound of the gallbladder–biliary tree.
4. Percutaneous transhepatic cholangiography.
5. Endoscopic retrograde cholangiopancreatography.

Oral cholecystography is still a valuable procedure despite its introduction to medicine in 1924. The majority of biliary calculi contain less than 4% calcium and require contrast agents to make them visible roentgenographically. Only 10 to 20% of biliary calculi are radiopaque. Iopanoic acid, the contrast agent, must be secreted and conjugated by the liver for the gallbladder to be visualized. Although cystic duct obstruction will not allow visualization of the gallbladder with iopanoic acid, incomplete cystic duct obstruction will allow visualization in the presence of disease.

Iopanoic Acid Visualization of the Gallbladder in the Presence of Disease

Finding	Implication
1. Calcification of the gallbladder wall (porcelain gallbladder)	Seen in chronic cholecystitis: 25% go on to develop carcinoma of the gallbladder
2. Opaque bile gravitates to dependent portion of the gallbladder (milk of calcium bile)	Cystic duct obstruction
3. Air in the gallbladder (emphysematous cholecystitis)	Choledochoduodenal or jejunal fistula secondary to choledocholithiasis with perforation
4. Poor opacification with augmented opacification of stones	Rapidly acting cholecystographic agent (noted when using ipodate calcium)
5. Layering of calcium on roentgenogram beam	Specific gravity of bile increases with contrast, causing cholesterol calculi to float; nonopaque colored stones sink
6. Persistent dense opacification at 24–36 hr	Acalculous cholecystitis
7. Calculi identified after fat ingestion	Excess contrast material in gallbladder obscures tiny calculi; fatty meal induces cholecystokinin secretion which contracts gallbladder
8. Adenomatosis (intramural diverticuli)	Predisposes to gallbladder calculi

Other findings that may be visualized with iopanoic acid:

1. Inflammatory polyps.
2. Cholesterol polyps: macrophages filled with cholesterol mimicking calculi.
3. Adenomatosis (intramural diverticuli): irregular contour of the gallbladder wall.
4. Overlying intestinal gas; simulates gallbladder defects.
5. Enterohepatic recirculation of conjugated iopanoic acid gives persistent dense opacification.

Nonvisualization of the Gallbladder

Causes for Nonvisualization of the Gallbladder with Iopanoic Acid when Disease Is Present

1. Reabsorption of iopanoic acid in the setting of an inflamed gallbladder (acute cholecystitis)
2. Cystic duct obstruction resulting in the small bowel being coated with contrast agent; conjugated iopanoic acid, when excreted by the liver, bypasses the gallbladder and enters the small bowel directly

Causes for Nonvisualization of the Gallbladder with Iopanoic Acid when Disease Is Absent

1. Malabsorption
2. Pancreatic insufficiency
3. Diet low in fat
4. Recent abdominal surgery
5. Peritonitis
6. Hepatic disease: bilirubin greater than 1.8 mg/dl may result in inadequate conjugation

Hepatocholescintography is the best method to differentiate acute from chronic cholecystitis. The radionuclide used, iminodiacetic acid (IDA), has several advantages over iopanoic acid cholecystography:

1. Gas and feces do not obscure scintiscan.
2. Pancreatitis does not affect scintiscan unless there is concomitant cholecystitis.
3. Scintiscan is not limited by modest elevation of bilirubin (bilirubin 4–8 mg/dl); IDA is metabolized by hepatocytes and excreted unconjugated in the hepatic ducts.

Increased liver impairment causes decreased bile flow and decreased uptake of nuclide by the gallbladder. Four patterns of radionuclide excretion are listed.

Patterns of IDA Secretion				
Condition	Liver	Gallbladder	Common bile duct	Duodenum
Normal gallbladder function	5 min	30–45 min	45–60 min	60 min
Acute cholecystitis	5 min	Not seen	45–60 min	60 min
Chronic cholecystitis	5 min	2–2.5 hr (not seen at 45 min)	45–60 min	60 min
Common bile duct obstruction	5 min	30–45 min and 2–2.5 hr	45–60 min	Not seen

Ultrasound of the gallbladder–biliary tree gives anatomic but not physiological information. It is valuable for its simplicity and can be performed at the patient's bedside. In addition, it has the ability to "view" organs adjacent to the biliary tree. Radionuclide hepatocholescintography can follow the ultrasound study to establish the physiological functioning of the biliary system. Dilated biliary hepatic radicles and/or common bile duct may not occur in acute choledocholithiasis and may mislead the clinician.

Criteria for Acute Cholecystitis Based on Sonographic Analysis

1. Gallbladder thickening greater than or equal to 5 mm
2. Gallbladder anechoicity (single or multilayered continuous or focally interrupted bands of edema, blood, or cell infiltrate perimuscularly; liver edema may also contribute to this sonographic feature)
3. Gallbladder distention; external anterior–posterior width greater or equal to 4 cm

There is a correlation between the pathological severity of inflammation and the degree of sonographic wall thickening and aneochoicity. Acute gangrenous cholecystitis can not be distinguished from nongangrenous cholecystitis by this diagnostic procedure.

Differential Diagnosis of Ultrasound Findings

Wall Thickness

1. Ascites as a result of congestive heart failure or cirrhosis

(continued)

Differential Diagnosis of Ultrasound Findings (*Continued*)

Wall Thickness (*Continued*)

2. Hepatitis
3. Acute and chronic cholecystitis
4. Adenomyomatosis
5. Gallbladder neoplasms
6. Renal failure
7. Multiple myeloma
8. Physiological contraction of the gallbladder
9. Artifactual transducer placement and angulation not along true long axis of the gallbladder

Gallbladder Distention

1. Diabetes
2. Postvagotomy
3. Fasting postoperatively
4. Narcotic analgesics
5. Watery diarrhea; hypokalemia–achlorhydria syndrome

Delineation of the Cystic Duct

1. Problem of fibrous spiral valves in the neck of the gallbladder
2. Tortuous cystic duct and surrounding structures may cause acoustic interference

Percutaneous transhepatic cholangiography is a roentgeno-graphic technique for visualizing the liver and the entire length of the common bile duct. The invasive radiologist usually must puncture the liver several times percutaneously under fluoroscopy before an appropriately sized target duct is located that will allow the passage of guidewires and catheters into the distal common bile duct. Prior review of bleeding parameters, antibiotic prophylaxis, analgesic/sedative premedication, sterile precautions, and informed consent are all necessary. This procedure does not visualize the gallbladder or bypass cystic duct obstruction. Biliary drainage can be done either internally or externally; percutaneous catheters are distally looped in the duodenum. In the majority of cases, the catheter can be passed beyond the site of biliary obstruction.

Endoscopic retrograde cholangiopancreatography (ERCP) is a procedure in which a catheter is inserted via an endoscope through the proximal alimentary tract into the ampulla of Vater.

The sphincter of Oddi is then entered, and contrast material is used
to outline the common bile duct and/or pancreatic duct. The ERCP
allows visualization of gallbladder stones 3–7 mm in size by means
of retrograde filling of the gallbladder and distal biliary tree.

C. Treatment of Symptomatic Biliary Tract Disease

In past years the only treatment of biliary tract disease
involved rest, nothing by mouth, anticholingergic medications,
and/or surgery. In the medically unstable elderly patient, cho-
lecystostomy was often the operation of choice unless there was
cystic or common bile duct obstruction. Choledochoduodenos-
tomy/jejunostomy was done only if distal common bile duct
obstruction could not be relieved by T-tube drainage.

As in the past, acute illness must be treated aggressively with
the patient being placed NPO (fluid and electrolyte balance to be
managed intravenously); antibiotic coverage should be initiated, and
pain medications given as tolerated. Since morphine constricts the
biliary tree, meperidine (Demerol®) may be a better choice. This
agent should be reserved only for cases of severe pain, however;
constipation, CNS toxicity, and orthostatic hypotension are major
limiting factors. If symptoms persist, additional intervention must
be initiated.

1. Current Treatment for Symptomatic Biliary Tract Disease

Present treatments	Indications
Percutaneous transhepatic biliary drainage (PTHBD)	Bypass common bile duct obstruction with cholangitis/hepatic abscess(es)
	Bypass solitary common bile duct stone
	Bypass common bile duct stricture
	Extract or crush common bile duct stone
Endoscopic retrograde cholangiopancreatography (ERCP)	Extract stone in a basket or crush common bile duct stone

(*continued*)

Present treatments	Indications
	Papillotomy and stone extrusion
	Insert stent for stricture
Cholecystostomy	Medically unstable patient with acute abdomen or sepsis attributable to cholecystitis
	Marked scarring resulting in gallbladder adhering to liver, thus making cholecystectomy very difficult
Cholecystectomy	Gallbladder empyema
	Gallbladder gangrene
	Gallbladder perforation

2. Treatment of Asymptomatic Biliary Tract Disease

In most cases no treatment is necessary in the older patient with asymptomatic biliary disease, i.e., stones in the gallbladder. The older patient with recurrent episodes of cholecystitis or who is diabetic, however, should have elective surgery if he or she is considered to be a candidate, as they are at high risk for recurrence.

X. PRESENTATION OF ACUTE ABDOMEN IN THE ELDERLY

Atypical presentation of catastrophic events within the gastrointestinal tract is a major cause of increased morbidity and mortality in the elderly. The physician must have a high degree of suspicion to avoid delay in making a diagnosis and initiating treatment. Supportive therapy often is the mainstay of treatment until definite therapy can be initiated. Early surgical consultation is necessary in most cases.

Predisposing Illnesses Capable of Causing an Acute Abdomen in the Elderly

1. Acute cholecystitis
2. Chronic cholecystitis with cystic duct obstruction
3. Choledocholithiasis

(continued)

Predisposing Illnesses Capable of Causing an Acute Abdomen in the Elderly (*Continued*)

    4. Small bowel obstruction
       a. Hernias (midline, ventral, femoral, inguinal)
       b. Adhesions and bands from previous surgeries
       c. Mesenteric thrombosis
    5. Large bowel obstruction
       a. Cecal or sigmoid volvulus
       b. Neoplasms
       c. Constipation with fecal impaction
       d. Medication-induced bowel atony
    6. Penetrating or perforating peptic ulcer
    7. Diverticulitis
    8. Appendicitis
    9. Pancreatitis

A. Evaluating an Elderly Patient with Acute Abdomen

Clinical examination of the abdomen may not correspond to the patient's poor clinical status. Although anorexia, nausea, and abdominal distention may be the only symptoms, atypical presentation may include confusion, dyspnea, and nonspecific deterioration in patient's condition. In the elderly the characteristic features of the disease causing "acute abdomen" are often minimal or absent. Raised white cell count, pyrexia, and increased pulse rate, for example, may not be seen in cases of cholecystitis, appendicitis, and diverticulitis. In a patient with an acute abdomen, laboratory evaluation should normally include the following:

    1. Complete blood count.
    2. Serum electrolytes.
    3. Serum osmolality $= [2\times (Na^+ + K^+)] + \dfrac{glucose}{18} + \dfrac{BUN}{2.8}$
    4. Blood sugar.
    5. Serum acetone.
    6. Serum amylase.
    7. Liver function tests.
    8. BUN and creatinine.
    9. Urinalysis.
   10. Chest X ray.
   11. Abdominal X ray.
   12. ECG.

Minimal peritoneal signs and ileus may confuse the clinician. Chemical peritonitis precedes bacterial peritonitis, influencing the time course of the illness. Caution is advised not to misinterpret palpation of a tortuous abdominal aorta, a finding common in the elderly, as a dissecting abdominal aortic aneurysm.

On an abdominal flat plate X ray, "bulging" or various aspects of the abdomen may assist in diagnosis. Asymmetric "bulging" of the flanks suggests contiguous inflammatory or mass lesion of the intraperitoneal, retroperitoneal, or subhepatic area. Since the liver is usually protected from extrahepatic inflammation by its peritoneal covering, it may be used to compare density of nearby accumulated fluids. Because the spleen has a noncontiguous peritoneal covering, it is a less reliable roentgenographic marker. Symmetrical bulging of the flanks on X-ray usually suggests an intraperitoneal mass and/or the presence of ascites.

B. Treatment and/or Other Diagnostic Procedures

Exploratory laparotomy is indicated in a patient with rapid deterioration. Definitive treatment and rapid resolution of the problem are necessary in most cases, and exploratory laparotomy may be lifesaving in a patient with rapid deterioration who may otherwise become inoperable if a decision is delayed. Other diagnostic procedures and therapeutic interventions are available. The following summarizes some of those more commonly used:

Procedure	Indication
1. Nasogastric tube suctioning with daily abdominal flat-plate roentgenogram	Toxic megacolon from ulcerative colitis or perforated diverticulitis
2. Upper gastrointestinal endoscopy	Gastric outlet obstruction; Malory–Weiss tear
3. Barium enema	Sigmoid/cecal volvulus
4. Lower gastrointestinal endoscopy	Ischemic/ulcerative colitis; diverticulitis; volvulus; Ogilvie's syndrome (spastic ileus secondary to malignant invasion of the celiac plexus)
5. Peritoneoscopy (laparoscopy)	Biliary obstruction; peritoneal/hepatic metastasis Shrunken or dilated gallbladder; fat necrosis in acute pancreatitis

(continued)

Procedure	Indication
6. Peritoneal lavage or tap	Mesenteric infarction; perforation; ascites; intraperitoneal bleeding
7. Angiography	Superior mesenteric artery thrombosis
8. Percutaneous transhepatic cholangiography and biliary drainage	Biliary obstruction

SUGGESTED READING

Almy T. P.: Facfors leading to digestive disorders in the elderly. *Bull NY Acad Med* 57:709, 1981.

Almy T. P., Howell D. A.: Diverticular disease of the colon. *N Engl J Med* 302:324, 1980.

Brocklehurst J. C., Khan Y.: A study of faecal stasis in old age and use of dorbanex in its prevention. *Gerontol Clin* 11:293–300, 1969.

Brocklehurst J. C., Kirkland J. L., Martin J., et al: Constipation in long stay elderly patients: Its treatment and prevention by lactulose, poloxalkol, dihydroxyanthroquinolone and phosphate enemas. *Gerontology* 29:181–184, 1983.

Brodribb A. J.: Treatment of symptomatic diverticular disease with a high-fibre diet. *Lancet* 1:664, 1977.

Burakoff R.: An updated look at diverticular disease. *Geriatrics* 32:83–91, 1981.

Cohen N.: Gastroenterology in the aged. *Mt Sinai J Med* 47:142–149, 1980.

Dodds W. J., Hogan W. J., Helm J. F., et al: Pathogenesis of reflux esophagitis. *Gastroenterology* 81:376–394, 1981.

Eastwood H. D. H.: Bowel transit studies in the elderly. Radioopaque markers in the investigation of constipation. *Gerontol Clin* 14:154–159, 1972.

James O.: Gastrointestinal emergencies in the elderly, in Coakley D. (ed): *Acute Geriatric Medicine.* London, Croom Helm, 1981, p 234.

Martin F., Farley A., Gagnon M., et al: Comparison of the healing capacities of sucralfate and cimetidine in the short-term treatment of duodenal ulcer: A double blind randomised trial. *Gastroenterology* 82:401–405, 1982.

Milne J. S., Williamson J.: Bowel habit in older people. *Gerontol Clin* 14:56–60, 1972.

Peterson B. H., Kennedy B. J., Butler R. N., et al: Aging and cancer management, *CA* 29:322–340, 1979.

Smith R. G., Rowe M. J., Smith A. N., et al: A study of bulking agents in elderly patients. *Age Ageing* 9:267–271, 1980.

Wormsley K. G.: Duodenal ulcer: An update. *Mt. Sinai J Med* 48:391–396, 1981.

5

Rheumatologic Problems in the Elderly

I. INTRODUCTION

Pain and limited mobility are common problems resulting from disorders of the musculoskeletal system in the elderly.

Since many factors may be involved, a thorough evaluation for any presenting joint and/or muscular problem is essential. Although osteoarthritis is the most common diagnosis made, problems such as sepsis, rheumatoid arthritis, gout, systemic lupus erythematosis, and malignancy often lead to confusion. In addition to certain diseases having an increased prevalence with age, normal age-related changes may predispose the older individual to problems. This chapter reviews the normal aging changes as well as the more common collagen vascular and degenerative illnesses associated with advancing age. The most important therapeutic goal in any of these illnesses is early recognition and initiation of therapy in order that maximal function and independent life style can be maintained.

Musculoskeletal Changes Associated with "Normal" Aging

1. Decrease in the intervertebral spaces leading to a loss of height
2. Formation of osteophytes on vertebrae
3. Fibrillar degeneration of the superficial cartilage of the weight bearing joints
4. Decrease in lean body mass
5. Decrease in bone mineralization

II. JOINT DISORDERS

Elderly persons commonly complain of joint pain from a variety of causes. Early evaluation and treatment are essential in order that further damage of the joint be prevented. In addition, failure to treat joint problems promptly may lead to the rapid loss of mobility associated with the joint pain.

Classification of Joint Disorders in the Elderly

1. Degenerative joint disease (osteoarthritis)
   a. Primary
   b. Secondary
      1. Prior trauma
      2. Prior inflammatory disease
      3. Metabolic joint disease
      4. Prior bone or cartilage disorders
      5. Inherited or developmental bone or cartilage disorders
2. Crystalline arthritis
   a. Gout (primary or secondary)
   b. Pseudogout
3. Septic arthritis
4. Rheumatic disease
   a. Rheumatoid arthritis
   b. Gout
   c. Systemic lupus erythematosus
5. Tumor-associated arthropathy

A. Osteoarthritis

Osteoarthritis (OA), or degenerative joint disease, is the number-one crippling disorder among older adults. Approximately 40 million Americans, including 85% of those over the age of 65, exhibit some radiologic evidence of OA; approximately 5 million of this group are symptomatic with pain, limitation of motion, joint stiffness, and swelling. The disorder is characterized pathologically by a deterioration of articular cartilage as well as a change in existing subchondral and even new bone formation.

Radiographic Features of Osteoarthritis

1. Asymmetric focal degenerative changes
2. Reactive subchondral bone formation
3. Joint space narrowing
4. Osteophytes at bony margins
5. Subchondral cysts
6. Chondrocalcinosis

Osteoarthritis is primarily a disease of cartilage and as such is not a classic inflammatory joint disease such as rheumatoid arthritis and other inflammatory joint diseases.

Inflammatory versus Noninflammatory Arthritis

Inflammatory	Noninflammatory (OA)
1. Joint pain and stiffness	1. Joint pain and stiffness
2. Constitutional symptoms	2. Absence of constitutional symptoms
3. Morning stiffness	3. Low-grade morning stiffness resolving rapidly
4. Mild to moderate activity decreases stiffness	4. Activity increases symptoms
5. Inflammatory signs prominent	5. Minimal inflammatory signs
6. Soft tissue swelling	6. Joint swelling as a result of bony enlargement

Although the etiology of OA is poorly understood, initial changes can be found in the cartilage, and this soon progresses to involve subchondral bone and synovial membrane. These pathological changes clinically manifest with pain and limited mobility. The pain tends to worsen with increasing use of the affected joint, and at times joint instability is noted. Physically, the joint is usually tender to palpation, with pain and crepitations often noted on motion. Weight-bearing joints are most often affected; pain and stiffness routinely worsen with increased use during the course of a day. The diagnosis of OA can be divided into primary and secondary presentations.

Classification of Osteoarthritis

Primary

1. Localized (usually weight-bearing joints)
2. Diffuse idiopathic skeletal hyperostosis (DISH, Forestrier's disease)

Secondary

1. Trauma
2. Prior inflammatory joint disease
3. Prior bone disorders
4. Metabolic disorders
5. Neuropathic arthropathy
6. Hemophiliac arthropathy
7. Congenital disorders

Joints Most Commonly Affected by Osteoarthritis

1. Distal interphalangeal joints (Heberden's nodes, if present, are often a sign of genetic predisposition)
2. Proximal interphalangeal joints
3. Carpal, metacarpal
4. Knees (often related to obesity, especially with medial compartment involvement)
5. Hips
6. Cervical spine
7. Lumbar spine
8. Metatarsalphalangeal (joint of great toe)

1. Treatment of Osteoarthritis in the Elderly

Treatment can be divided into four phases. All patients should be taught how to care for involved joints properly in order to preserve maximal functioning. A referral to a skilled physical therapist may be necessary to help with patient education as well as to develop a specifically designed program of rehabilitation. Medications such as aspirin and the nonsteroidal antiinflammatory agents are often helpful in reducing pain. Divided doses should be given throughout the day to insure adequate blood levels. Titration of medications starting with a low dose will prevent many side effects from interfering with continued treatment. In the case of aspirin, a small increase in dose may cause a marked increase in serum salicylate levels because of enzyme saturation. The presence of tinnitus in a patient is often a warning sign of toxicity requiring a reduction in aspirin dosage. In some selected cases, intraarticular

steroid injections may be necessary if more conservative management has failed. In all cases, a septic joint must be excluded prior
to initiating this form of treatment. When the affected joint has
progressed to the point that mobility is threatened, surgical replacement of the joint (usually hip) should be considered.

2. Management of Osteoarthritis

1. Patient education	Active participation is needed
2. Physical therapy	Joint protection exercises Joint rest
3. Drug therapy	Analgesics, aspirin Nonsteroidal antiinflam- matory therapy ? Intraarticular steroid injection
4. Surgical	Joint replacement

B. Septic Arthritis

At one time, septic arthritis was considered an illness that
primarily affected children. Since 1960, however, the incidence has
been rising in the elderly, particularly with the advent of joint
replacement surgery. Classically, the illness presents as severe pain
in the involved joint that is exacerbated with movement. Since
elderly persons often present with an atypical clinical presentation,
a delay in diagnosis is not uncommon.

Differential Diagnosis of an Inflamed Joint in the Elderly

1. Rheumatoid arthritis
2. Gout
3. Pseudogout
4. An acute flare of osteoarthritis
5. Tumor-induced arthropathy
6. Cellulitis

Sepsis may also be superimposed on a preexisting problem such as rheumatoid arthritis, an osteoarthritic joint that has received intraarticular injections of steroids, or a prosthetic joint. Radiologic changes may not occur at an early stage; the absence of radiographic findings should not rule out the diagnosis. Aspiration of the joint fluid with gram stain and culture is the diagnostic method of choice. Although synovial white blood cell counts are usually markedly increased, the elevation is only diagnostic if above 100,000/mm^3. In the case of cellulitis, extreme caution must be exercised when aspirating a joint to avoid inoculating a previously sterile joint with microorganisms.

Staphlyococcus aureus is the most common organism cultured from the elderly person's infected joint. Gonococcal arthritis is more common in a younger population. A variety of organisms have been cultured from affected joints. These include streptococci, pneumococci, and gram-negative bacilli. Unusual organisms can be cultured from a previously damaged joint or from an immunosuppressed patient. Tuberculous arthritis, previously thought to be limited to children and young adults, is now seen frequently in the elderly. It usually presents as a chronic monoarticular arthritis, most commonly affecting the vertebral column and large weight-bearing joints.

Treatment mandates repeated aspiration to drain the infected joint of destructive enzymes. Secondly, the joint should be placed at rest. Finally, a course of parenteral antibiotics directed toward the causative bacterial organism should be initiated. In the case of tuberculous arthritis, active tubercular therapy should be initiated.

C. Rheumatoid Arthritis

Rheumatoid arthritis (RA) is a systemic disease that is often associated with other immunologic illnesses and syndromes. Although normally considered a chronic debilitating disease of the young, its highest incidence and prevalence are found in the elderly. Although it is not surprising that the prevalence of this chronic illness increases with age, it is noteworthy that the incidence also increases with age.

The diagnosis of RA in the elderly may be obscured by its insidious onset. In addition, joint pains associated with degenerative joint disease may add to the confusion.

Symptoms of Rheumatoid Arthritis in the Elderly

1. Morning stiffness
2. Joint tenderness or pain on motion
3. Soft tissue swelling of one joint
4. Soft tissue swelling of second joint within 3 months
5. Soft tissue swelling of symmetrical joints
6. Subcutaneous nodules

The most important characteristics in the elderly include joint pain, swelling, and tenderness. In addition, morning stiffness lasting longer than 30 min is an important distinguishing feature and may often be the earliest diagnostic finding. Constitutional symptoms, although often present, are not specific and could indicate the presence of other underlying illness.

Associated laboratory findings may include an anemia of chronic disease, elevated erythrocyte sedimentation rate (ESR), and a positive rheumatoid factor; none of these values are diagnostic. Anemia is an age-prevalent illness with multiple etiologies; other causes of anemia should be ruled out before one concludes that the anemia is a manifestation of RA. Likewise, the ESR is often mildly elevated in the elderly and as such is neither sensitive nor specific in the diagnosis of RA. Finally, the rheumatoid factor may be positive in as many as 10–20% of elderly persons over the age of 65, usually with titers less than 1 : 160. Although the presence of high titers of rheumatoid factor may be suggestive, it is not diagnostic. The diagnosis of RA is a complex one made largely on the basis of clinical presentation.

Older persons with a longstanding diagnosis of RA may have variable presentations. The spectrum of disease ranges from inactive with bony deformities and rheumatoid nodules signifying earlier active disease to a full range of articular and systemic manifestations that are symptoms of ongoing active disease. Initial presentation in the elderly patient may be either acute or insidious. An insidious onset is usually associated with a better prognosis. In general, elderly persons have fewer systemic manifestations. The shoulder and cervical spine are the most commonly involved joints. Occasionally, a more acute onset with rapid manifestations of arthritic and constitutional symptoms may occur. Acute-onset RA has been noted in elderly patients who present with a GI malignancy. In the absence of malignancy, acute RA may often undergo clinical remission.

Treatment for RA follows the same course as for OA. The first phase is patient education. Patients should be told what to expect during the course of the illness and how to best maintain normal functioning. Although rest is an important component of therapy, it is essential that the older individual maintain a base-line activity level. Prolonged bed rest may result in an overall decline in function, reduced muscle mass, negative calcium balance, and an increased susceptibility to infections. Referral to a physical therapist, exercise training, and the judicious application of local heat may preempt the need for prolonged use of medications.

Salicylate therapy with careful monitoring of serum levels should be used. The elderly often have problems with gastrointestinal side effects and tinnitus, requiring initial treatment with low doses to be titrated up slowly. Alternatively, nonsteroidal anti-inflammatory medications (NSAIDs) may be used; as with all medications, care must be exercised, and all potential side effects must be considered. If the patient has a previously recognized aspirin hypersensitivity, the NSAIDs are contraindicated as well. Renal insufficiency is a significant risk; potassium-sparing diuretics and NSAIDs have a synergistic effect on reducing renal function.

More aggressive therapy including the use of gold salts, systemic steroids administration, antimalarial agents, penicillamine, and low-dose methotrexate are controversial in elderly persons with RA and should be used cautiously if at all. The toxic effects of these medications should be weighted heavily against any possible therapeutic benefit.

D. Gout

Gout is a disease that primarily affects middle-aged and elderly men (90% of all cases). Onset is usually noted in the fourth and fifth decades of life, and both the incidence and prevalence increase with age. Hyperuricemia often results in monosodium urate monohydrate crystal deposition in the articular joints because of the relatively poor solubility of uric acid. In addition, approximately 20% of those with gout develop nephrolithiasis as a complication. The arthritis associated with the crystal deposition may be isolated to several acute episodes lasting several days, or it may progress and become recurrent with joint destruction and the appearance of tophi. The initiation of therapy with multiple medications, however, has greatly reduced the incidence of crippling arthritis. Hyperuricemia results from either an overproduction or an impaired renal secretion (seen in up to 90% of affected individuals) of uric acid. Malignancy can result in secondary elevations in uric acid through overproduction. The symptoms of acute gouty arthritis are those of an acute painful arthritis with joint inflammation and abrupt onset. Symptoms usually peak within 24–48 hr and subside in several days. More severe attacks, however, may last up to several weeks.

Joints Commonly Associated with Acute Gouty Arthritis Attacks

Most Common
1. First metatarsal phalangeal joint; may be involved in up to 75% of affected individuals
2. Ankle joints
3. Knee joints

Less Common
1. Finger joints
2. Wrist joints
3. Elbow joints

Uncommon
1. Shoulder joints
2. Hip joints
3. Sacroiliac joints

Diagnosis can be confirmed by analyzing fluid of the affected joint under a polarizing light microscope. The presence of strongly negative birefringent crystals under polarized light is diagnostic of gouty arthritis. Occasionally, gouty nephropathy results. This is usually manifested by proteinuria, decreased creatinine clearance, and azotemia. Fortunately, this is not a common manifestation of the disease.

The chronic course of the disease is manifest by the presence of tophi in multiple areas of the body, usually occurring approximately 10 years after the initial presentation of the disease. Since asymptomatic hyperuricemia is clearly associated with the risk of

developing arthritic and renal complications, treatment should be
initiated prior to symptomatology developing.

Chronic Tophaceous Gout: Areas of Tophi Distribution

1. Synovium
2. Subchondral bone
3. Olecranon bursa
4. Infrapatellar region
5. Achilles' tendon
6. Subcutaneous tissues of forearm and overlying joints
7. Helix of ear

Therapy of gout should be aimed at treating the acute gouty
arthritis with prevention of further recurrence. The level of serum
uric acid must be reduced.

Treatment of Gout

Acute Attacks	Comments
1. Nonsteroidal antiinflammatory agents	Pain from acute attacks is particularly sensitive to indomethacin, 50 mg q.i.d. for 4–7 days; gastrointestinal and mental status side effects, however, may be limiting factors in the elderly
2. Colchicine	0.6 mg every hour until pain relief or development of nausea and/or diarrhea occurs; no more than 12 tablets/day may be ingested; more toxic in elderly population; dehydration and electrolyte disturbances must be carefully observed

(continued)

Treatment of Gout (*Continued*)

Acute Attacks	Comments
3. Phenylbutazone	100 mg q.i.d.; short-term administration may be acceptable; usually not recommended in elderly patients
4. Corticosteroids	Rarely recommended; may be helpful if side effects of other medications are prohibitive or if attack is unresponsive to other medications

Chronic Recurrent Gouty Arthritis	Comments
1. Prophylaxis	Colchicine, 0.6 mg, 2–3 tablets daily, will often prevent recurrence of acute attacks; may be necessary if recurrent attacks are common
2. Probenecid	1 mg q.d. in divided doses promotes uric acid excretion; side effects: minimal fever, rash, G.I. upset
3. Sulfinpyrazone	800 mg/day is maximum dose; promotes uric acid secretion; side effects minimal: fever, rash, G.I. upset
4. Allopurinol (Zyloprim®)	300–800 mg/day; inhibits xanthine oxidase; side effects include drug rash, transient leukopenia, transient liver function abnormalities; useful therapy in most cases of asymptomatic hyperuricemia

E. Calcium Pyrophosphate Dihydrate Crystal Deposition Disease (Pseudogout)

Pseudogout (CPPD) is an inflammatory joint disease that affects both middle-aged and elderly individuals. Clinically, it can be divided into six separate patterns.

Calcium Pyrophosphate Dihydrate Crystal Deposition Disease: Clinical Types

1. Pseudogout: 10–20% of patients; knee is the most commonly affected joint
2. Pseudorheumatoid arthritis: 2–6% of patients; continuous attack of arthritis characterized by morning stiffness
3. Osteoarthritis without acute pseudogout: 35–60% of patients; typically this is a bilateral and symmetrical arthritis of a chronic and progressive nature
4. Osteoarthritis with acute attacks of pseudogout: 10–35%; clinically resembles osteoarthritis with superimposed acute episodes
5. Asymptomatic chondrocalcinosis: often found by radiologist on routine X ray; most often noted in knee, symphysis pubis, and wrist films
6. Pseudo-Charcot joint: up to 2% of patients; present in the absence of neurological disease

---

As with all inflamed joints, the diagnosis is made by microscopic examination of the affected joint's synovial fluid. Classically, crystals weakly positive under a polarized light microscope are seen in the absence of bacteria. A septic joint may be superimposed on any inflamed joint and must be ruled out prior to a diagnosis of CPPD. When the diagnosis of CPPD is made, a list of associated disorders should be considered.

Disorders Associated with CPPD

1. Hyperparathyroidism
2. Hemochromatosis
3. Wilson's disease

(continued)

Disorders Associated with CPPD
(Continued)

4. Hypophosphatasia
5. Hypomagnesemia
6. Ochronosis
7. Hypothyroidism
8. Diabetes mellitus
9. Gout
10. Osteoarthritis

Hyperparathyroidism and hemochromatosis, in particular, should be excluded when the diagnosis of CPPD is made.

Pathologically, calcium pyrophosphate crystals are deposited within the different articular structures. The cause of this deposition is unclear. The presence of these crystals within the articular structures causes both acute and chronic inflammatory changes.

Radiologically, the diagnosis of CPPD may be made by examining X-rays of the knee, symphysis pubis, and wrist. If the knee films alone are examined, approximately 90% of patients will be found positive. Addition of symphysis pubis films increases the sensitivity to 98%. If all three films are obtained, the detection rate is close to 100%.

Treatment for CPPD depends on the clinical manifestation of the problem. Although asymptomatic chondrocalcinosis diagnosed by X-ray often requires no treatment whatsoever, the more inflammatory presentations of the disease may require acute interventions with antiinflammatory medications such as aspirin, the NSAIDs,

F. Differential Laboratory Characteristics of Joint Disease

Laboratory	OA	RA	Trauma	Sepsis	Crystalline arthritis
WBC	Normal	Variable	Normal	Increase	Increase
ESR	Normal	Increase	Variable	Increase	Increase
Synovial fluid					
Appearance	Clear	Turbid	Clear, bloody	Purulent	Turbid, purulent
Viscosity	High	Low	High	Low	Low
WBC (per mm³)	2,000	2,000	2,000	50,000	2,000
Glucose	Normal	Decrease	Normal	Decrease	Normal
Microscopic	No cells	Cells	Cells	Cells	Crystals under polarized light

and even colchicine, as in the case of pseudogout. As previously mentioned, the use of colchicine in the elderly requires careful titration because of decreased tolerance of the gastrointestional side effects.

G. Commonly Used Anti-inflammatory Medications in the Elderly

Medication	Plasma half-life	Dosage range (mg/day)	Side effects	Excretion	Comments
1. Aspirin	20 min 3–6 hr	2500–7500	GI upset; CNS; tinnitus; renal	Renal	Therapy should be followed with serum levels
Propionic acid series					
2. Ibuprofen	2 hr	1200–2400	GI upset; rash; headache; blurred vision; fluid retention; renal insufficiency	Renal	Currently available in OTC preparations
3. Naproxen (Naprosyn®)	13 hr	500–750	GI upset; rash; headache; tinnitus; edema; renal insufficiency	Renal	
Indolelike series					
4. Sulindac (Clinoril®)	8 hr	150–400	GI upset; rash; dizziness; tinnitus; edema; renal insufficiency	Renal	
5. Tolmetin sodium (Tolectin®)	1 hr	600–1800	GI upset; rash; dizziness; tinnitus; edema; renal insufficiency	Renal	
6. Indomethacin (Indocin®)	4.5 hr	25–800	GI upset; mental confusion; headache; visual disturbance; renal insufficiency	Renal	May be useful in cases of acute gouty arthritis

III. SYSTEMIC LUPUS ERYTHEMATOSUS

Approximately 12% of all new cases of systemic lupus erythematosus (SLE) occur in the elderly. The presentation of the illness is often atypical and usually follows a more benign clinical course.

Criteria for Diagnosing Systemic Lupus Erythematosus

1. Malar rash that tends to spare the nasolabial folds
2. Discoid rash
3. Photosensitivity
4. Oral ulcers

(continued)

Criteria for Diagnosing Systemic Lupus Erythematosus (*Continued*)

5. Arthritis affecting two or more peripheral joints
6. Serositis–pleuritis or pericarditis
7. Renal disorder: persistent proteinuria or cellular casts
8. Neurological disorder: seizures or psychosis
9. Hematological disorder: hemolytic anemia, leukopenia, lymphopenia, or thrombocytopenia
10. Immunologic disorder: positive LE preparation, anti-DNA to native DNA, anti-Sm, or false positive serologic test for syphilis known to be positive for at least 6 months
11. Antinuclear antibody (ANA) in the absence of drugs that are known to cause a "lupus syndrome"

At least four of the criteria need to be present either serially or concurrently in order to make the diagnosis. Because the normal initial manifestations of SLE in the elderly include arthritis, rash, and constitutional symptoms (low-grade fever, malaise, weight loss), the diagnosis is often delayed. Patients may not seek medical attention until the occurrence of more debilitating symptoms such as pleuritis or pericarditis. Unfortunately, SLE is often not considered as part of the differential diagnosis in the elderly. Other frequent clinical manifestations of SLE in the elderly include nephritis, neuropsychiatric sequelae, alopecia, hepatomegaly, and splenomegaly.

Clinical Features of SLE in the Elderly

1. Arthritis (most common)
2. Cutaneous signs
3. Nephritis
4. Pleurisy
5. Pericarditis
6. Parenchymal pulmonary abnormalities
7. Neuropsychiatric effects
8. Alopecia
9. Hepatic disorders
10. Splenomegaly
11. Nasopharyngeal ulcers
12. Adenopathy
13. Raynaud's phenomenon (least common)

The laboratory evaluation of SLE in the elderly is similar to that required in the younger population.

Common Laboratory Features of SLE in the Elderly

1. Anemia
2. Leukopenia
3. Lymphopenia
4. Thrombocytopenia
5. Antinuclear antibody (positive)
6. Lupus erythematosus preparation (positive)
7. Anti-DNA antibodies (positive)
8. Rheumatoid factor (positive)
9. ESR > 50 mm/hr (Westergren)

The prognosis of SLE in the elderly is better than that of a younger population. If corticosteroid therapy is required, the use of smaller doses may be sufficient to treat the more malignant components of the disease. As in the case of all medication use in the elderly, careful titration is required to minimize side effects.

IV. POLYMYALGIA RHEUMATICA AND TEMPORAL ARTERITIS

Polymyalgia rheumatica (PMR) and temporal arteritis (TA) are the second most common (after rheumatoid arthritis) inflammatory rheumatologic disorders in the elderly. Although the two disorders are related, the exact association currently remains unclear. Because of the often nonspecific presentation of illness, a high index of suspicion is necessary in order to accurately assess and diagnose these disorders. Delay of diagnosis and treatment may lead to permanent impairment, particularly in the case of TA, where blindness may result.

Characteristics of Polymyalgia Rheumatica

1. Primarily affects elderly white women with an annual incidence of 50/100,000 population in all adults over 50 years of age

---

(continued)

Characteristics of Polymyalgia Rheumatica
(Continued)

2. Primary complaints include early morning muscle stiffness preferentially affecting the shoulder, proximal arm, hip, and proximal thigh muscle groups
3. Systemic manifestations include malaise, weight loss, low-grade fever, anemia, and an elevated ESR
4. Abrupt onset is common

The diagnosis is made on the basis of the clinical presentation. The laboratory evaluations are nonspecific; the ESR, however, is usually elevated greater than 50 mm/hr (Westergren).

Therefore, the presence of morning stiffness with preserved muscle strength, systemic symptoms, and an elevated ESR should lead to a high degree of suspicion. In most cases, the presence of a normal ESR rules out the diagnosis; any preexisting condition capable of lowering the ESR, however, must be excluded.

Temporal arteritis is primarily a disease of elderly white men and women with a median age of 65–75 years. Onset is usually insidious, occurring over a 1- to 3-month period. Onset of complaints is usually nonspecific.

Symptoms of Temporal Arteritis

Symptom	Comment
1. Headache	Often unilateral over the temporal region
2. Nonspecific myalgia or arthralgia	May be a manifestation of coexisting PMR; must be differentiated from other rheumatologic disorders
3. Jaw claudication	Often the cardinal symptom that may increase the index of suspicion for TA
4. Visual changes	Nonspecific complaints may precede permanent blindness
5. Systemic complaints	Anorexia; fever; weight loss; malaise

The presence of associated symptoms particularly related to vision plus a very high ESR, often exceeding 100 mm/hr (Westergren), is usually considered sufficient clinical evidence to initiate treatment with corticosteroids prior to biopsying the temporal artery. In order to confirm the diagnosis of TA, temporal artery biopsy is essential; biopsy must preceed initiation of corticosteroids to maximize diagnostic acumen. A wide specimen of the temporal artery is necessary for diagnosis, as the pathological changes associated with giant cell arteritis are not present in all areas of the artery. If the first biopsy is negative and a high index of suspicion still remains, biopsy of the contralateral temporal artery may be indicated. Although the use of corticosteroids may change the results of the biopsy, treatment should not be withheld pending biopsy if it is not readily available. Delay in treatment may result in permanent blindness.

Pathological Changes Seen in Temporal Artery Biopsy

1. Panarteritis with polycellular infiltrates
2. Segmental involvement: "skip lesions"
3. Giant cells in internal elastic layer
4. Intimal proliferation
5. Absence of fibrinoid necrosis

The two conditions may coexist; studies suggest that 50% of patients with PMR will also have coexistent TA and 50% of patients with TA will have coexistent PMR. Some authors have recommended temporal artery biopsy in all patients with the diagnosis of PMR; the efficacy of this approach, however, remains controversial.

An accurate diagnosis of TA is essential because of the higher dosage of corticosteroids required to control the symptoms of TA and prevent blindness.

Treatment of TA and PMR

Temporal Arteritis: Sequential Steps

1. Prednisone 60 mg/day in divided dosage for 5 days
2. Prednisone 60 mg/day in single dose for at least 4 weeks

(continued)

Treatment of TA and PMR (Continued)

Temporal Arteritis: Sequential Steps (Continued)

3. Slow and gradual tapering of no more than 5 mg/week down to 20 mg/day in a single dose
4. Slower tapering of 1 mg/day per week of therapy; maintenance dose of therapy of 5 mg/day may be necessary for 2–3 years
5. Follow course of therapy with serial evaluation of symptoms and ESR
6. If ESR climbs or symptoms reoccur, repeat course of therapy

Polymyalgia Rheumatica

1. Rule out TA
2. Prednisone 20 mg/day
3. Taper rapidly to 5 mg/day maintenance dose
4. Maintenance therapy should be continued for 1–2 years
5. Nonsteroidal antiinflammatory medications may be sufficient treatment when maintenance dose of prednisone has been attained

Prednisone therapy is required for the treatment of these disorders. As in many other conditions, the risk–benefit ratio of prolonged corticosteroid therapy must be evaluated. The use of long-term high-dose prednisone therapy may result in severe complications for the elderly patient.

Complications of Prolonged Prednisone Therapy in the Elderly

1. Osteoporosis
2. Aseptic necrosis of bone
3. Immunosuppression
4. Myopathy
5. Pancreatitis
6. Delirium
7. Glaucoma
8. Edema
9. Hypokalemia
10. Diabetes mellitus

(continued)

**Complications of Prolonged Prednisone Therapy in
the Elderly** (*Continued*)

11. Hyperlipidemia
12. Cushingoid features
13. Impaired wound healing
14. Depression
15. Hemorrhagic gastritis
16. Pseudotumor cerebri
17. Suppression of hypothalamic–pituitary–adrenal axis

V. AMYLOIDOSIS

Amyloidosis is a spectrum of diseases in which insoluble
proteinaceous material is deposited in several different organ systems. The deposition of this material may result in severe impairment of the affected system, often leading to death. It is primarily a
disease of the elderly and is often associated with an underlying
systemic illness, most commonly multiple myeloma.

Classification of Amyloidosis

Condition	Comment
1. Primary generalized amyloidosis	No concurrent associated disease; amyloid deposition found in heart, tongue, carpal tunnel, gastrointestinal tract, peripheral nerves, skin, joint, and skeletal muscle
2. Secondary amyloidosis	Usually associated with chronic inflammatory or chronic infectious states; deposition found in liver, kidney, spleen, and adrenal gland
3. Multiple-myeloma-associated amyloidosis	Same organ distribution as primary amyloidosis

---

(*continued*)

Classification of Amyloidosis (*Continued*)

Condition	Comment
4. Heredofamilial amyloidosis	Seen in many genetically related illnesses, particularly mediterranean familial fever; amyloid deposition in nerves, heart, and kidney
5. Local amyloidosis	Manifested by local tissue deposition; rare
6. Senile amyloidosis	Usually benign infiltration of many tissues

Symptoms of amyloidosis are nonspecific since the proteinaceous deposits affect multiple organ systems. Diagnosis is based on a high clinical index of suspicion and biopsy of the specific organ involved when possible. In most cases, a rectal or gingival biopsy is sufficient to confirm the diagnosis. The biopsy specimen when stained with Congo red has a pathognomonic green birefringence under polarized light microscopy.

The disease is usually fatal, with death occurring from cardiac arrhythmia or renal failure when those organs are involved. There are no specific treatments for amyloidosis, and therapy must be directed toward any underlying reversible etiology. Since amyloid deposition is found in many organ systems on autopsy of the elderly, the term amyloidosis is reserved for those cases in which amyloid deposition has resulted in organ impairment with clinical signs and symptoms.

VI. POLYMYOSITIS AND DERMATOMYOSITIS

Polymyositis and dermatomyositis are related disorders that occur commonly in the elderly.

Polymyositis is an inflammatory disorder of the skeletal muscle, whereas dermatomyositis is clinically similar with the addition of a characteristic skin rash. It is primarily a disease affecting the young (5–15 years old) and those in their middle and later years (greater than 45 years old).

Criteria for Diagnosis of Polymyositis and Dermatomyositis

1. Symmetrical proximal muscle weakness
2. Elevated serum muscle enzymes
3. Myopathic EMG
4. Muscle biopsy abnormality

The diagnosis of polymyositis is complete if all four criteria are met. The diagnosis of dermatomyositis is complete if three of the criteria plus a characteristic skin rash are identified.

Clinical Features of Polymyositis/Dermatomyositis

Feature	Comment
1. Proximal muscle weakness	Insidious onset (3–6 months) marked by exacerbations and remissions
2. Muscle atrophy and tenderness	
3. Contractures	
4. Heliotrope rash	A lilac discoloration of upper eyelids
5. Periorbital edema	
6. Gottron's sign	Scaling erythematous eruption over knuckles, elbows, knees, medial malleoli, neck, face, and upper chest
7. Periungual erythema	Desquamation of fingertips
8. Sclerodactyly	
9. Raynaud's phenomenon	
10. Arthralgias	
11. Systemic	Fever; anorexia; weight loss; fatigue; malaise
12. Cardiac	Arrhythmias; congestive failure
13. Pulmonary	Interstitial lung disease

Laboratory Features of Polymyositis/Dermatomyositis

1. Elevation of serum muscle enzymes, CPK, and aldolase in 98% of patients
2. ESR variable; not specific
3. Anemia: normochromic, normocytic; usually mild

The EMG generally has a classic triad of findings, including (1) small-amplitude, short-duration, polyphasic motor potentials, (2) fibrillations with increased insertional activity, and (3) spontaneous high-frequency discharges.

The presence of the clinical symptoms, laboratory abnormalities of elevated serum muscle enzymes, and typical EMG findings confirm the diagnosis.

Treatment is initially conservative, progressing to a more aggressive approach as the illness worsens.

Treatment Strategies in Polymyositis/Dermatomyositis

1. Early bedrest
2. Mobilization and physical therapy
3. Patient education
4. Prednisone, similar dosage as used in TA; maintenance dose is usually 20 mg/day for at least 6 months, sometimes for as long as 5 years; no proven efficacy
5. Immunosuppressive agents if treatment fails with prednisone

Patients should be followed closely in order to evaluate therapeutic efficacy. Measurement of serum CPK and muscle strength are thought to correlate well with disease activity. A questionable association exists between the presence of polymyositis/dermatomyositis and malignancy. This remains an area of controversy and requires further evaluation. All patients with the diagnosis of polymyositis/dermatomyositis should have a careful history and physical examination looking for signs and symptoms suggestive of malignant illness. If any such sign or symptom is found, the diagnosis of malignancy must be vigorously pursued;

however, extensive evaluation in the absence of any clinically observed signs or symptoms is probably not warranted.

VII. VASCULITIS

The term vasculitis refers to a spectrum of disorders associated with inflammation and necrosis of blood vessels. The disorders range from a mild, secondary component of a primary disease to a more malignant group of primary disorders. The underlying etiology of each of the vasculitides is felt to be mediated by an immune mechanism; however, further work is required to better define the specific pathogenesis.

Classification of the Vasculitides Commonly Seen in the Elderly

Classification	Common
1. Polyarteritis nodosa (PAN)	Characterized by sub-cutaneous nodules, ulceration, peripheral gangrene, and vascular mottling; may respond to corticosteroids
2. Allergic granulomatosis (Churg–Strauss syndrome)	Clinically similar to PAN; characterized further by eosinophilia, history of allergies, and lung involvement
3. Serum sickness	Hypersensitivity vasculitis; therapy consists of removal of offending antigen
4. Henoch–Schonlein purpura	Not usually seen in elderly; associated with IgA deposition in blood vessels; hypersensitivity reaction usually following upper respiratory viral illness
5. Wegener's granulomatosis	Uncommon illness manifested by granulomatous involvement of upper

(continued)

Classification of the Vasculitides Commonly Seen in the Elderly (Continued)

Classification	Common
	and lower respiratory tract; glomerulonephritis also occurs; treatment consists of cyclophosphamide
6. Giant-cell arteritis	See section on PMR and TA
7. Erythema nodosum	Painful nodular inflammatory process of the dermis and subcutaneous tissues
8. Essential mixed cryoglobulinemia	Represents a specific entity of IgM rheumatoid factor directed against the IgG molecule

The above listing accounts for a small subset of the whole spectrum of vasculitis. The delineation and treatment of the whole spectrum of disorders go beyond the scope of this handbook. The most clinically relevant vasculitis in the elderly is giant-cell arteritis, reviewed in Section IV. Hypersensitivity vasculitis, associated with the use of medications, is also commonly seen in the elderly.

Common Medications Associated with Hypersensitivity Vasculitis in the Elderly

1. Penicillin
2. Sulfonamides
3. Phenobarbital
4. Phenothiazines
5. Phenytoin sodium (Dilantin®)
6. Thiazide diuretics
7. Furosemide (Lasix®)
8. Quinidine

SUGGESTED READING

Baer A. N., Pincus T.: Occult systemic lupus erythematosus in elderly men. *JAMA* 24:3350–3352, 1983.

Blau S. P.: Polymyalgia rheumatica: A syndrome associated with many different diseases. *Geriatr Med Today* 3(10):79–84, 1984.

Bohan A., Peter J. B.: Polymyositis and dermatomyositis. *N Engl J Med* 292:344, 1975.

Calkins E., Challa H. R.: Disorders of the joints and connective tissue, in Andres R., Bierman E. L., & Hazzard W. R. (eds): *Principles of Geriatric Medicine* New York, McGraw-Hill, 1985, pp 813–843.

Clough J. D., Elrazak M., Calabrese L. H.: Weighted criteria for the diagnosis of systemic lupus erythematosus. *Arch Intern Med* 144:281–285, 1984.

Coles L. S., Fries J. F., Kraines R. G. et al: From experiment to experience: Side effects of nonsteroidal anti-inflammatory drugs. *Am J Med* 74:820–828, 1983.

Fauci A.: The spectrum of vasculitis. *Ann Intern Med* 89:660–676, 1978.

Fauci A. S.: Corticosteroids in autoimmune disease. *Hosp Pract* 18(19):99–114, 1983.

Grinblat J.: Senile amyloidosis, *Pract Gastroenterol* 7(6):36–52, 1983.

Henriksson K. G.: Polymyositis and dermatomyositis: Association with malignancy in older patients. *Geriatr. Med. Today* 2(12):86–96, 1983.

Hochberg M. C.: Osteoarthritis: Pathophysiology, clinical features management. *Hosp Pract* 19(12):41–53, 1984

Manchui L. A., Jin A., Pritchard K. I.: The frequency of malignant neoplasms in patients with polymyositis–dermatomyositis. *Arch Intern Med* 145(10):1835–1839, 1985.

Miller M. L., Kantrowitz F. G., Campion E. W.: Rheumatology in geriatrics, in Gambert S. R. (ed): *Contemporary Geriatric Medicine*, (Vol. 1.) New York, Plenum Press, 1983, pp 155–203.

Porta-Sales J, Ferrer-Ruscalleda F.: Temporal arteritis: Present concepts *Cardiovasc Rev Rep* 6(3):310–321, 1985.

Rodnan G. P., Schumacher H. R. (eds), Zvaifler, N.J. (assoc ed): *Primer on the Rheumatic Diseases*. Atlanta, Arthritis Foundation, 1983.

Rossman I.: The pathology of normal aging. *Geriatr Med Today* 3(4):37–43, 1984.

Terkeltaub R.: Rheumatoid arthritis of late onset. *Geriatr Med Today* 4(6):101–110, 1985.

6

Genitourinary Problems in the Elderly

I. NORMAL AGE-RELATED CHANGES

"Normal" age-related changes affecting the kidney result in a decrease in renal function of approximately 0.6% per year of life following maturity. Despite this change, serum creatinine varies little, if any, with age (normal range 0.8–1.2 mg/dl). This is because the older person has less muscle mass and, therefore, less creatinine formed. Although difficulties in collection may preclude its usefulness, a creatine clearance determination based on a 24-hr urine collection is the most accurate way to assess renal function in the elderly.

The following formula can be used to estimate creatinine clearance as a function of age:

$$\text{Creatinine clearance* } (C_{Cr})\text{men} = \frac{(140 - \text{age}) \times (\text{weight in kg})}{(72 \times \text{serum creatinine})}$$

It is important to remember that although increasing age, in itself, results in a decline in renal function, past and/or current diseases affecting the kidney will further compromise creatinine clearance. This has significant implications when prescribing medications that are cleared through the kidney or considering certain dietary supplements.

*Multiply by 0.85 to calculate values for women. From Gault M. H., Cockcroft D. O.: Creatinine clearance and age. *Lancet* 2:612, 1975.

Age-Related Changes in Renal Function

1. Kidneys decrease approximately 20% in size by age 70
2. Number of glomeruli decreases by 30–50% by age 70
3. Glomerular basement membranes thicken
4. Tubular surface area decreases
5. Renal loss is more pronounced among the juxta-medullary nephrons
6. Decrease in urinary concentrating ability
7. Creatinine clearance declines approximately 30% from the fourth to eighth decade of life
8. Renal blood flow decreases by 50%
9. Decreased ability to maintain acid–base equilibrium

II. ACUTE RENAL FAILURE

Acute renal failure (ARF) can be classified as either being prerenal, postrenal, or parenchyrenal in origin. Acute renal failure is a rapid deterioration of glomerular filtration resulting in the accumulation of nitrogenous wastes.

Causes of Acute Renal Failure

Prerenal

1. Decreased perfusion (hypotension, poor cardiac output)
2. Extracellular volume depletion (blood loss; diuretics; dehydration; vomiting; nasogastric suctioning)

Renal

1. Acute tubular necrosis (nephrotoxins; prerenal causes; transfusion reactions)
2. Vasculitis
3. Uric acid nephropathy
4. Glomerulonephritis
5. Hypertension
6. Hypercalcemia
7. Hepatorenal syndrome

Postrenal

1. Bladder outlet obstruction (prostatic hypertrophy; tumor)
2. Ureteral obstruction (adhesions; strictures; clots; calculi; extrinsic compression)

A. Prerenal Azotemia

Prerenal azotemia results from an underperfusion of the renal arterioles. Increased sodium retention may result as a compensatory mechanism to help expand effective circulating blood volume. In most cases, this results in a low urinary concentration of sodium. Other causes of decreased effective renal artery perfusion include either an expanded (congestive heart failure) or contracted (dehydration, nephrotic syndrome, hepatorenal syndrome) extracellular fluid volume.

Classically, urine osmolality exceeds serum osmolality. Increased reabsorption of urea nitrogen results in a higher BUN in relation to serum creatinine; the BUN : creatinine ratio is usually greater than 10. Decreased renal perfusion may lead to increased renin release; secondary aldosteronism may result in a high serum potassium.

B. Treatment of Prerenal Azotemia

Specific therapy must be directed at the underlying causative factor. Fluid deficits must be corrected. Although a fluid challenge with saline or IV furosemide may help differentiate this problem from other causes of renal failure, extreme caution and close monitoring are advised.

Hypotension and congestive heart failure can also lead to a hypoperfusion of the kidney, and treatment must be aimed at correcting the underlying problem. A decrease in free water clearance may potentially lead to hyponatremia as well as congestive heart failure. In these cases, fluid restriction may be necessary despite low renal perfusion. In the elderly, careful monitoring is an essential component of care, and there should be a relatively low threshold for transfer to a "monitored bed."

C. Acute Parenchymal Renal Failure

Acute parenchymal renal failure (APRF) is associated with significant morbidity and mortality in the elderly. The clinical course varies greatly because of the wide variety of precipitating factors. Diagnosis may be delayed until late in the course of illness because of underreporting of illness, poor history, or confusion as to etiology.

Causes of APRF

1. Acute tubular necrosis (ATN)
   a. Volume contraction secondary to hypotension
   b. Low-cardiac-output state
   c. Hemorrhage
   d. Toxin
   e. Hemolysis
   f. Rhabdomyolysis
2. Toxin-induced renal failure
3. Glomerulonephritis
4. Renal artery occlusion
5. Hypertension
6. Any disease of the intrarenal vasculature

Presentation of APRF

1. Impaired urinary concentrating ability
2. Decreased reabsorption of sodium
3. Urine and plasma osmolality approximately equal
4. Urine sodium concentration <40 mEq/liter
5. Urine may contain casts or renal tubular epithelial cells

Microscopic Evaluation of Urine

Microscopic Review of Urine	Diagnosis Suggested
RBC casts	Glomerulitis
Dark glomerular casts and epithelial cells	Acute tubular necrosis
Hyaline or finely granular casts	Prerenal azotemia
No significant casts or cells observed	Obstruction

Phases of Acute Tubular Necrosis

Oliguric Phase

Lasts days to weeks (rarely exceeds 1 month)
The 24-hr urine volume is usually between 50 and 400
ml/day (anuria is rare)

(continued)

Phases of Acute Tubular Necrosis (*Continued*)

Oliguric Phase (*Continued*)

Ends with a gradual increase in urine volume as a result
of diuresis of accumulated edema fluid

Diuretic Phase

At times diuresis may require large amounts of replacement fluid; monitoring is required of fluid and
electrolytes

Recovery Phase

Return of glomerular function may lag behind diuretic
phase
BUN and creatine should return to normal within several
days, with BUN the last to return to normal

D. Guidelines for Treating Postrenal (Obstructive) Uropathy

1. Should be considered in all cases of acute renal failure with
   reduced urinary output.
2. Early diagnosis is essential; permanent impairment in renal
   function can develop within as little as 7 days.
3. An intravenous pyelogram (IVP) is diagnostic; however, it
   should be done with caution and only after hydration; renal
   visualization can usually be obtained despite a serum
   creatine as high as 7 mg/dl even in the elderly.
4. A urinary catheter should be inserted to rule out lower tract
   urinary obstruction, e.g., prostatic enlargement.
5. Failure to visualize the renal pelvis and ureters on IVP
   indicates need for bilateral retrograde pyelography.
6. A postobstructive diuresis may follow the removal of the
   urinary tract obstruction; this results from an osmotic diuresis caused by accumulated urea nitrogen and a water
   diuresis clearing accumulated edema fluid.
7. Impaired tubular reabsorption of salt and water may occur.
   Caution is advised in fluid and electrolyte replacement. In
   general, fluids can be replaced based on 80–90% of the
   previous hour's urine output (0.45 normal saline is usually
   an acceptable starting fluid with changes made according to
   specific urinary electrolyte output).

E. Managing the Elderly Patient with Acute Renal Failure

1. Confirm the diagnosis.
2. Protect one arm from all blood drawing, IV, etc. in case vascular access for hemodialysis is eventually required.
3. Dialysis (peritoneal or hemodialysis) should be performed prior to the onset of uremia. Hemodialysis is preferred in cases of extensive tissue destruction, e.g., trauma, postoperative.
4. Restrict potassium and protein intake unless hemodialysis is being performed.
5. Avoid overhydration; replace GI, insensible, and urinary losses plus approximately 10 ml/kg per day.
6. Weigh patient daily to assess level of hydration.
7. Monitor fluid status very closely.
8. Urinary sodium loss should be measured and replaced. Potassium should be monitored and never given (even in medications, e.g., penicillin) unless patient becomes hypokalemic, in which case replacement should be limited to identifiable losses.
9. Hyperkalemia mandates immediate treatment.
   a. Sodium polystyrene sulfonate (Kayexolate®) is an exchange resin that binds K^+ in the GI tract and can be given orally or as an enema, 15–60 g in 3 to 4 ml/g fluid q.i.d. Sorbitol (20%) should be administered concomitantly to avoid the constipation associated with Kayexolate®, a problem particularly troublesome to the elderly.
   b. Glucose/insulin therapy is advised in conjunction with Kayexolate® when serum K^+ exceeds 6.5 mEq/liter. A rapid i.v. push of 25 g glucose and 8 to 10 units regular insulin shifts K^+ from the extracellular to intracellular space, lowering serum K^+ rapidly. This therapy is often combined with i.v. administration of $NaHCO_3$ (45 mEq).
   c. If K^+ exceeds 7.5 mEq/liter or ECG changes are noted, i.e., peak elevation of the T wave, prolonged Q–T interval, QRS widening, and prolonged PR interval, i.v. calcium (5 to 10 ml of 10% $CaCl_2$) should be given slowly to antagonize the effect K^+ has on the

myocardium. Repeat treatment every 30 min to 2 hr as
required.

10. Reduce dietary phosphorus intake; phosphate-binding ant-
    acids may be a useful adjunct therapy to keep PO_4^{3-}
    within normal limits.

11. Reduce oral intake of magnesium, primarily contained in
    antacid/laxative preparations.

12. Monitor patient's neurological status closely. Delirium,
    hallucinations, tetany, and frontal lobe depression and
    even convulsions may appear. Metabolic causes must be
    excluded. Uremic convulsions can be acutely treated with
    diazepam (5–20 mg i.v.) as needed over 3 to 5 min with
    close attention to respiratory status and providing ven-
    tilatory assistance as necessary. Diphenylhydantoin (Dilan-
    tin®) can also be given i.v.; however, 100 mg 2–4 times a
    day p.o. is preferred for the patient not in status epilep-
    ticus. Phenobarbital, 60–180 mg/day may be useful; how-
    ever, many elderly respond paradoxically to this
    medication. If all of the above fail to control the uremic
    convulsions, an i.v. lidocaine (10 mg) bolus followed by a
    continuous infusion of 30 μg/kg per min is usually
    effective.

13. Support patient as necessary until renal function returns to
    normal.

III. CHRONIC RENAL FAILURE

Patients with chronic renal failure (CRF) are living increasingly longer as a result of better medical management of this
complex problem. Although mortality in the first year on maintenance dialysis approaches 20%, this drops to 10% annually thereafter. In the United States, Medicaid pays for all treatments
required for end-stage renal disease. There are currently 30,000
patients on maintenance hemodialysis, with many more receiving
peritonal dialysis either at home or in dialysis centers. Renal
transplantation remains the preferred method for treating young
patients with CRF, with approximately 3000 transplants performed
annually in the United States. Fewer than 10% of all renal
transplants are performed in patients over 50 years of age despite
the fact that this older population accounts for 80% of all cases of

CRF. Although age does not appear to affect the way in which renal transplants function, the elderly tend not to tolerate the required immunosuppressive therapy as well.

Guidelines for Managing Chronic Renal Failure

1. Medical management should be continued until creatine clearance falls to about 10 ml/min
2. Dialysis should be initiated as soon as symptoms of uremia begin to appear, i.e., fatigue, nausea, vomiting. and/or neurological changes. Many advocate starting dialysis in the elderly prior to the appearance of symptoms
3. Once a decision has been made to initiate dialysis, the treatment most suitable for the patient must be chosen. Hemodialysis can be done either in a dialysis center or at home. Older patients, particularly those with coincident diabetes or hypertension, tend to do better in a dialysis center. Peritoneal dialysis is increasingly popular, particularly in the elderly patient suffering from diabetes, cardiovascular disease, or difficult vascular access. Although peritoneal dialysis is approximately 15–25% slower in its ability to clear toxic wastes from the blood, fewer side effects are noted, particularly in the elderly. Progression of diabetic retinopathy also appears to be slower with this method of dialysis. In general, continuous irrigation of the peritoneal membrane is required for approximately 40 hr per week. This usually is performed in two dialysis sessions per week at a center or in four overnight sessions at home. A minimal requirement for home dialysis includes the ability to perform the appropriate connection under sterile conditions
4. Chronic dialysis may lead to psychological problems: depression, poor self-image, and altered life style. Patients with CRF and their families may require psychological counseling.

IV. URINARY INCONTINENCE IN THE ELDERLY

Urinary incontinence is a common and frequently debilitating problem affecting elderly women more often than men. Although approximately 15% of elderly persons residing in the community have this problem, it has been estimated that as many as 25 to 35% of hospitalized and 40 to 60% of institutionalized elderly are affected.

Potential Complications of Urinary Incontinence

Increased risk of urinary tract infections
Increased use of antibiotics with potential side effects
Decubitus ulcers
Increased medical nursing time
Depression
Social isolation
Increased laundry costs
Increased risk of institutionalization

Mechanism of Urine Control

1. Urine flows into the bladder at a steady rate
2. Collected urine stimulates stretch receptors in the "bladder center" of the spinal cord ($S_{2,3,4}$)
3. A "bladder-filling" center (located in anterior portion of the cingulate and part of the superior frontal gyrus) indicates when the bladder is filled and voiding is necessary.
4. A "bladder inhibitory" center (located in the frontal cortex) develops during childhood to control motor activity of perineal structures helping to inhibit the sacral reflex action if desirable
5. When the bladder contraction generates pressure exceeding outlet resistance in the bladder neck, urethra, and external spincter, voiding occurs.

A. Urinary Incontinence: Failure to Store Urine

Any defect in the mechanism responsible for the normal flow of urine can result in urinary incontinence.

Causes of Urinary Incontinence in the Elderly

1. Fecal impaction
2. Hypercalcemia/hyperglycemia
3. Medications (diuretics, theophylline, etc.)
4. Coffee/tea/alcohol use
5. Pseudodementia
6. Attention-seeking behavior
7. Environment (poor access to bathroom/commode)
8. Urinary tract infections
9. Senile vaginal changes (trigonitis)
10. Pelvic floor incompetency
11. Neurogenic bladder
12. Bladder outlet obstruction

B. Evaluating the Elderly Patient with Urinary Incontinence

1. Comprehensive history and physical examination, including a pelvic and rectal examination.
2. Urinanalysis with urine culture if indicated.
3. Blood chemistry tests.
4. "Straight" urinary catheterization to assess degree of urinary retention.
5. Cystometrogram.

See table on pp. 132–133.

C. Nonpharmacological Treatment of Established Urinary Incontinence

1. Start only after all readily "treatable" and/or transient causes of urinary incontinence have been excluded.
2. Initiate an individualized bladder training program. This should include a toileting regimen and teaching maneuvers that either reflexly and/or mechanically facilitate bladder emptying: bladder reflex-triggering stimuli such as pinching the abdomen, suprapubic tapping, and running water; abdominal straining; external pressure applied to the lower abdomen (Credé maneuver).
3. Avoid medications/habits that may precipitate incontinence, i.e., diuretic use and coffee or tea consumption.

4. Avoid using "adult pads" indiscriminately, as these may in themselves stimulate a reflex bladder contraction.
5. Refer the patient with refractory bladder outlet resistance or continued urine leakage for further urological evaluation. A variety of surgical procedures, i.e., external sphinctero-tomy, surgical widening of the bladder neck, or trans-urethral resection of the bladder neck, may be of value in a given patient. An artificial sphincter may also prove effective in managing incontinence in selected cases.

D. Pharmacological Treatment of Established Urinary Incontinence

Numerous medications are available to help inhibit bladder contractility or increase outlet resistance. Side effects are not insignificant, and extreme caution is advised.

1. Propantheline (Probanthine®), is an anticholinergic agent that can help decrease bladder contractility. Resistance to medication action occurs over a period of months; increasing dosages may lead to debilitating side effects such as blurred vision, dry mouth, constipation, decreased diaphoresis, cardiac arrhythmias, and altered mentation.
2. Imipramine hydrochloride (Tofranil®) stimulates both the β-adrenergic receptors in the bladder body that inhibit contractility and the α-adrenergic receptors in the bladder base that increase outlet resistance. Numerous side effects (dry mouth, blurred vision, cardiac arrhythmias, and excessive sweating) limit usefulness. Close observation and caution are advised, especially for those elderly with preexisting cardiovascular disease.
3. Oxybutynin chloride (Ditropan®) and flavoxate hydro-chloride (Urispas®) exert direct antispasmotic action on smooth muscle. Oxybutynin also has an anticholinergic action with limiting side effects.
4. Ephedrine, pseudoephedrine (Sudafed®), and phenylpropa-nolamine (Ornade®) stimulate the α-receptor, resulting in increased bladder outlet resistance.
5. Bethanechol chloride (Urecholine®) increases bladder contractility and decreases bladder outlet resistance. Long-term effectiveness is unpredictable; resistance to treatment and numerous side effects limit usefulness.

The Neurogenic and Obstructed Bladder: Characteristics and Presentations

	Uninhibited neurogenic bladder	Reflex neurogenic bladder	Atonic neurogenic bladder	Autonomous neurogenic bladder	Bladder outlet obstruction
Commonly associated with	Advanced age; stroke; altered mental status	Infants; spinal cord lesions; severe cerebral disease	Diabetes with peripheral neuropathy; tabes dorsalis; anticholinergic agents	Tumors of cauda equina; spina bifida	Prostatic hypertrophy in men; pelvic pathology in women; chronic fecal impaction
Etiology	Damaged inhibitory center	Destroyed pathways to and from regulatory centers in brain	Inhibited posterior sacral afferent fibers from bladder	Destroyed segments of spinal cord	Obstruction to bladder outlet
Sensation of bladder filling	Present	Absent	Absent	Absent	Variable
Ability to inhibit bladder contractions	Absent	Absent	Poor	Absent	Partially present
Cystometric evaluation	A. Small to moderate bladder	A. Small bladder volume	A. Large residual volume	A. Constant dribbling	A. Constant dribbling

B. Early urge to micturate	B. Spontaneous bladder contraction	B. Spontaneous ineffective contractions	B. Small amounts of urine voided frequently	B. Distended bladder
C. Urinary retention rare	C. Urinary retention rare	C. Low bladder pressure	C. Urinary retention of small amounts common	
		D. Constant dribbling as a result of overflow	D. Frequent ineffective bladder contractions	
			E. Normal or slightly decreased bladder capacity	
Treatment Bladder training	Bladder training pharmacological agents as tolerated	Bladder training; straight catheterization at regular intervals; pharmacological agents as tolerated; discontinue potentially causative medications	Bladder training; pharmacological agents as tolerated	Correct underlying abnormality

V. URINARY CATHETERIZATION

If control of urine remains a problem despite all attempts at treatment, urinary catheterization may be the only alternative for management. Catheters are most commonly indicated in cases of urinary retention with resultant overflow dribbling.

A. Intermittent Catheterization

1. A safe and effective method of draining the bladder of urine.
2. Although easily performed in most cases, this technique requires either a reliable, cooperative, and well-motivated patient or a competent and conscientious care giver.
3. Because catheter insertion is required every 4 to 6 hr to prevent the bladder from overdistending, many find this method too time consuming.
4. Sterile technique is required in the hospital setting; although still controversial, clean but unsterile technique has been reported by many to be a safe and effective method in the home.
5. Although low-dose, long-term antibiotics are usually recommended, their usefulness in controlling asymptomatic bacteriuria in these patients remains controversial.
6. Symptomatic urinary tract infections must be treated with a full course of antibiotics as determined by urine culture and sensitivity testing.

B. Indwelling Urinary Catheterization

1. Rarely indicated except in cases of either documented and progressive damage to the upper urinary tracts despite ongoing alternative treatment or incontinence in the bedridden patient with a decubitus ulcer.
2. Catheter-related complications may include bladder and

renal stones; perinephric, vesical, or urethral abscess; and renal failure.

3. Asymptomatic bacteria in catheterized urine are expected and warrant no treatment.
4. A clinical urinary tract infection requires immediate treatment. Initial antibiotic coverage should be based on institution-prevalent organisms; urine culture and sensitivity will determine antibiotic of choice.
5. A closed catheter system is essential.
6. A catheter must be secured to minimize movement.
7. Urine must always flow downward.
8. The catheter must be changed any time the system becomes contaminated.
9. Special training is required for all persons assuming responsibility for catheter care.
10. The urogenital area should be cleansed with an iodophor · preparation prior to catheter insertion.
11. Catheters should be changed every 4 to 6 weeks or more often as indicated.

VI. RENAL LITHIASIS

Kidney stones occur in approximately 90–200 persons per 100,000 population. It has been estimated that 12% of the population will have a kidney stone sometime during their life. Although most patients present with their first stone during early adult life, stones are not uncommon in the elderly. Management, however, may need to be modified in relation to overall medical status.

Composition of Stones	Percent
Calcium oxalate	80
Struvite	10
(magnesium ammonium phosphate hexahydrate)	
Uric acid	7
Other	3

Disorders Associated with Renal Stone Formation

Hyperparathyroidism
Idiopathic hypercalciuria
Excessively high protein intake
Distal renal tubular acidosis (RTA)
Chronic hypophosphatemia
Primary hyperoxaluria
Medullary sponge kidney
Urinary tract infection
Vitamin C use
Hyperuricosuria
Elevated urine pH
Sarcoidosis
Vitamin D intoxication
Hyperthyroidism
Wilson's disease
Multiple myeloma
Bone metastases
Hyperadrenalism
Paget's disease
Excessive calcium intake with hypercalciuria

A. Evaluation for Renal Lithiasis

1. Obtain accurate history, including family history, medication use, and diet.
2. Common symptoms to elicit: pain, fever, hematuria, nausea, or vomiting.
3. Laboratory testing: urinalysis, renal function testing, complete blood count, blood chemistry testing.
4. Intravenous pyelogram.
5. Stone analysis.

B. Treatment of Renal Lithiasis

1. Determine if at all possible the etiology of the stone.
2. Calculi less than 7 mm will usually pass spontaneously and require supportive treatment only. Over 7 mm, surgical removal of the stone or extracorporeal shock wave lithotripsy may be required.

3. Insure diuresis; titrate fluid given to avoid overhydration.
4. Control pain: narcotics may be required; side effects including orthostatic hypotension and constipation must be watched for closely. Indomethacin may be effective in decreasing renal colic.
5. Antiemetics given cautiously may provide patient comfort.
6. Maintain fluid intake at 2–3 liters per day as tolerated.
7. Treat any precipitating problem that is identified.

SUGGESTED READING

Batlle D. C., Sehy J. T., Roseman M. K., et al: Clinical and pathophysiologic spectrum of acquired distal renal tubular acidosis. *Kidney Int* 20:389–396, 1981.

Chaussy C., Schmiedt E., Jochan D., et al: Extracorporeal shock and wave lithotripsy for treatment of urolithiasis. *Urology* 23:59–66, 1984.

Drach G. W.: Evaluation of the urinary stone former. *Semin Urol* 2:12–19, 1984.

Garibaldi R. A., Brodine S., Matsumiya R. N.: Infections among patients in nursing homes. *N Engl J Med* 305:731–735, 1981.

Johnson E. T.: The condom catheter: Urinary tract infection and other complications. *South Med J* 76:579–582, 1983.

Pletka P.: *Nephrolithiasis in Pathophysiology of Renal Disease.* New York, McGraw-Hill, 1981, pp 667–709.

Priefer B., Duthie E. H., Jr., Gambert S. R.: Frequency of urinary catheter change and clinical urinary tract infection in a skilled nursing home. *Urology* 20:141–142, 1982.

Rowe J. W., Andres R., Tobin J. D., et al: The effect of age on creatinine clearance in man: A cross-sectional and longitudinal study. *J Gerontol* 31:155–163, 1976.

Warren J. W., Muncie H. L., Jr., Bergquist E. J., et al: Sequelae and management of urinary infection in the patient requiring chronic catheterization. *J Urol* 125:1–8, 1981.

7

Endocrine and Metabolic Problems in the Elderly

I. OSTEOPOROSIS

Osteoporosis is a disorder characterized by lack of mineral in the axial skeleton as well as in the peripheral bones. Bones are structurally weak and easily prone to fracture. Approximately 200,000 hip fractures are diagnosed each year, primarily in elderly osteoporotic women.

Differential Diagnosis of Osteoporosis

Osteomalacia	Hyperthyroidism
Immobilization	Hyperadrenocorticism
Multiple myeloma	Heparin administration
Acromegaly	Chronic liver disease
Hyperparathyroidism	Chronic anemia

A. Treatment

It is essential that an acurate diagnosis of osteoporosis be made. Although specific therapy should be directed at disease-related complications, e.g., fractures and pain, life-long prevention is the hallmark of treatment.

In order to maintain calcium balance, premenopausal women should take at least 800 mg/day of elemental calcium. Postmenopausal women are advised to take up to 1500 mg of elemental calcium and 400 IU vitamin D per day under medical supervision. Periodic monitoring of 24-hr urinary excretion of calcium will provide early warning in the unlikely event of hypercalciuria.

Physical activity as tolerated should be encouraged. A number of agents are currently being assessed in regard to their efficacy in preventing osteoporosis, i.e., estrogens, anabolic steroids, calcitonin, vitamin D metabolites, parathyroid hormone, fluoride, and diphosphonates. Although studies using bone density have in many cases provided promising results, fracture prevention is the major goal. Estrogens (0.625 to 2.5 mg conjugated estrogens/day) either given cyclically or with premarin (10 mg days 21–25) appear to provide the best retardation of bone resorption. Elderly women on estrogen therapy, unless posthysterectomy, must be followed gynecologically because of a possibly increased risk of endometrial carcinoma. (See table on p. 141.)

B. Calcium Equivalents

Quantities of foods needed to supply the amount of elemental calcium (291 mg) in one cup (8 oz) of whole milk are listed.

Food Source	Approximate Measure
Roast beef	5 1b
Eggs	10
Peanut butter	29 tbsp
Salmon with bones and oil	4 oz
Sardines	2 ¼ oz
Tuna	7 ¾ lb
Rice	14 cups
Oatmeal	13 cups

(continued)

Distinguishing Features of Metabolic Bone Disease

Condition	Symptoms	Serum			Urine			Radiographic findings
		Calcium	Phosphorous	Alkaline phosphatase	Calcium	Phosphorous	Hydroxyproline	
Osteoporosis	Dowager's hump Vertebral compression fractures Hip or wrist fracture	Normal	Normal	Normal or elevated with fracture	Normal	Normal	Normal	Cortical thinning; resorption of endosteal bone; osteopenia; vertebral wedging with increased vertical trabeculae
Osteomalacia	Proximal muscle weakness Generalized bone pain	Normal or low	Normal or low	Elevated	Low	Elevated	Elevated	Osteopenia; pseudofractures
Paget's disease	Localized bone pain Bone deformity Neurological abnormalities	Normal	Normal	Elevated	Normal	Normal	Elevated	Osteopenia and/or sclerosis
Hyperparathyroidism	Renal lithiasis Gastrointestinal discomfort CNS disturbance	Elevated or normal	Low	Normal or increased	Elevated or low	Elevated	Elevated	Subperiosteal bone resorption; bone cysts; disappearance of lamina dura

Food Source	Approximate Measure
Bread (white) slices	13–15
Cornflakes	73 cups
Egg noodles	18 cups
Apples	29
Bananas	29
Corn	36 cups
Potatoes, baked	22
Greens (collards, kale, turnip, mustard)	1 cup
Broccoli	2 cups
Yogurt	1 cup
Pudding	1 cup
Milk, sweetened condensed	$\frac{1}{3}$ cup
Milk, evaporated	$\frac{1}{2}$ cup
Ice cream	$1\frac{3}{4}$ cup
Buttermilk	1 cup
Cheese (American, cheddar)	$1\frac{1}{2}$ oz
Cottage cheese (creamed and low fat)	2 cups
Cottage cheese (dry curd)	6 cups
Swiss cheese	1 oz
Cheese spread	2 oz

II. OSTEOMALACIA

Osteomalacia, the adult form of rickets, results from deficiency in vitamin D. Vitamin D plays a major role in promoting calcium absorption by the gut as well as calcification of osteoid in the bone. As serum calcium declines, and parathyroid hormone increases compensatorily, bone resorption takes place. This is coupled with a decline in phosphorus and an increase in calcium reabsorption in the kidney. Repletion of vitamin D corrects all pathological processes and lesions.

Although vitamin D deficiency has become extremely rare in young populations who consume fortified milk, osteomalacia is being diagnosed in the elderly with increasing frequency. Failure to consume fortified milk, avoidance of sun exposure, and a higher intake of medications that may interfere with vitamin D metabolism in the liver (diphenylhydantoin and/or phenobarbitol) or utilization (diphosphonates, cholestyramine, corticosteroids, and aluminum hydroxide gels) are contributing factors.

An accurate diagnosis can be facilitated by a thorough history and physical examination.

An elevation in serum alkaline phosphatase accompanied by a low level of serum calcium and phosphorus are pathognomonic. Levels of vitamin D and 25-hydroxyvitamin D are low; 1,25-dihydroxyvitamin D level may be low or normal. There will be a low urinary calcium excretion as well. Rarely will a bone biopsy be indicated. If done, specimens must be clearly labeled "bone biopsy: rule out osteomalacia" to avoid specimens being decalcified prior to analysis. Malabsorption of fat-soluble vitamins should be ruled out by history; definitive testing mandates a 72-hr stool collection for fecal fat.

Initial treatment is 800–2000 units of vitamin D per day for 2 to 3 weeks. Although rare, hypocalcemia resulting from "hungry bone syndrome," a deposition of calcium into depleted bone, should be watched for early in treatment. Thereafter, maintenance therapy with 400 units daily should be given. In cases of fat malabsorption, high doses of vitamin D will be necessary (50–500,000 units vitamin D daily). These high doses of vitamin D should be avoided unless specifically indicated, as they can further lead to demineralization. Vitamin D deficiency resulting from interfering medication use mandates a reevaluation of prescription practices.

III. PAGET'S DISEASE

Paget's disease is a focal disorder of unknown etiology characterized initially by excessive resorption and subsequently by excessive formation of bone. This may result in a "mosaic" pattern of lamellar bone associated with extensive local vascularity and increased fibrous tissue in the adjacent marrow.

Paget's disease begins during mid- to late life and has been reported in as many as 0.1 to 3.0% of populations studied. Most cases have been reported in the United Kingdom, Australia, New Zealand, Germany, France, and the United States. The disease is rare, however, in India, China, Japan, the Middle East, Africa, and Scandinavia.

Although a familial background of Paget's disease has been noted in as many as 20% of reported cases, Paget's disease is largely sporadic in nature and has no HL-A antigen preponderance. Men are affected more frequently than women (6:4).

A. Biochemical Features of Paget's Disease

1. No change in bone amino acid composition.
2. Increased collagen cross linkages in pagetic bone (hydroxylysine, norleucine).
3. Increased urinary excretion of hydroxyproline (reflects increased bone degradation).
4. Increased serum alkaline phosphatase (reflects increased osteoblastic activity).
5. Hyperuricemia common.

B. Metabolic Aspects of Paget's Disease

1. Serum Ca^{2+} and PO_4^{3-} normal.
2. Parathyroid hormone not usually affected.
3. Urinary calcium excretion not elevated in the absence of fractures and immobility.
4. Increased overall rate of bone mineral turnover.
5. Deposition and resorption of calcium may be increased in excess of 20-fold.

C. Clinical and Pathological Phases of Paget's Disease

1. **Osteolytic or destructive phase.** Increase in osteoclastic cell activity; cells often multinucleated and in abnormal shapes; rarely pure; osteoblastic response usual.
2. **Mixed phase.** Osteoblastic response results in deposition of lamellar bone adjacent to areas of irregular resorption or on opposing portions of the trabeculae. "Woven bone" common. Bone marrow may be replaced by loose fibrous connective tissue. Hypervascularity and fibroblastlike mesenchymal cells may be seen. Cement lines may be noted.
3. **Osteoblastic phase.** Increased bone deposition may be noted. Rarely seen in pure form.

D. Common Focal Manifestations

Spine
1. Pain from involvement at any site
2. Nerve root compression
3. Spinal cord compression: most common in thoracic area; rare in cervical spine
4. Spondylitislike syndromes

Skull
1. Osteoporosis circumscripta (osteolytic), usually painless
2. Cranial enlargement (mixed), may be painful
3. Basilar invagination, rarely results in neurological complications
4. Temporal bone involvement, may result in hearing loss
5. Ossicle involvement, hearing loss common
6. Trigeminal neuralgia
7. "Pagetic steal syndrome"

Facial and Jawbones
1. Maxillary involvement (leontiasis ossea)
2. Peridental involvement with cemental hyperplasia, loss of lamina dura, and displacement of teeth

Pelvis and Extremities
1. Pain may occur in any area
2. Hip involvement: acetabular and femoral head disease may result in protrusio acetabuli
3. Knee pain, may occur with distal femoral disease
4. Tibial and femoral bowing and pain common; humerus less common
5. Small bone involvement may occur; large bones affected more commonly

E. Local Complications

1. **Fractures,** usually transverse and perpendicular to cortex. Incomplete fractures may be precursor lesions, especially in femur and tibia. Fractures in pagetic bone usually heal well.

2. **Neoplasia.** Overall incidence of neoplasia is less than 1%, with osteosarcoma or giant-cell tumor most common. Although metastatic tumors are rarely reported in pagetic bone, they still must be considered. Chondrosarcomas, reticulum-cell sarcomas, and even multiple myeloma tumors have been reported.

F. Systemic Complications of Paget's Disease

1. **Hypercalciuria:** more common than hypercalcemia and may result in renal lithiasis.
2. **Hypercalcemia:** rarely occurs without extensive involvement, immobilization, and/or fractures.
3. **Hyperuricemia and gout:** commonly reported, particularly in men.
4. **Calcific periarteritis.**
5. **Cardiovascular.**
   a. High-output state. Although changes are usually apparent only after Paget's disease involves greater than 30% of the skeleton, significant increases in cardiac output may be seen with less extensive disease.
   b. Intracardiac calcification may be seen, particularly involving the aortic and mitral valves.
6. **Cutaneous vasodilation.**
7. **Angioid streaks.** Approximately 10% of advanced cases have these associated findings. Rarely, these are associated with pseudoxanthoma elastica.
8. **Malabsorption syndrome** (rare).

G. Criteria for Treatment

1. Disabling pain not relieved by analgesics and antiinflammatory agents.
2. Progression of skeletal disease, i.e., increasing deformity, bone enlargement, frequent fractures, vertebral compression, or acetabular protrusion.
3. Rapid decline in hearing.
4. Neurological complications.
5. Congestive heart failure caused by the high-output state.

6. Serum alkaline phosphatase and/or urinary hydroxyproline excretion four or more times above normal in symptomatic patients.

H. Therapeutic Modalities

1. **Analgesics** provide symptomatic relief of pain.
2. **Calcitonin.** Salmon calcitonin, an analogue of the human hormone, has been approved for treatment despite the fact that its therapeutic mechanism is unknown. Fifty Medical Research Council (MRC) units injected daily subcutaneously for 6 to 12 months may result in prolonged remission. Side effects may include transient nausea, vomiting, and flushing. Patients should be skin tested prior to use to exclude those allergic to this agent. Although most patients report relief of pain within 2–6 weeks of starting therapy, resolution of neurological complications is less common, and deafness is rarely improved. Levels of serum alkaline phosphatase and urinary hydroxyproline usually decline with drug therapy and can serve to indicate disease activity. Need for subcutaneous injections limits usefulness; many patients become refractory to this form of therapy.
3. **Diphosphonates.** This class of compounds inhibits osteoclastic bone resorption. Etidronate sodium (Didronel®) is approved by the FDA for the treatment of Paget's disease, and other compounds are currently under investigation. Etidronate sodium is given orally in a dose of 5 mg/kg 2 hr after breakfast. Although side effects are negligible at this dosage, only 6 months of treatment is recommended by the manufacturer. A remission frequently results; therapy should be repeated if the disease becomes active again. Osteomalacia has been reported as a side effect when diphosphonates are used in larger doses and for longer periods of time. Benefit of oral use is a great advantage.
4. **Combination of Calcitonin and Etidronate Sodium.** This combined therapy is particularly useful in cases of acute neurological complications. Side effects are similar to those when each therapy is used alone.
5. **Mithramycin.** This agent is rarely indicated because of its high side-effect profile. It is thought to treat Paget's disease

by damaging osteoblasts: a dose of 15 to 25 µg/kg is given parenterally. Extreme caution is advised when using this agent in the elderly patient.

IV. HYPERCALCEMIA IN THE ELDERLY

The patient can be diagnosed as having hypercalcemia if (1) the serum calcium concentration is more than 10.5 mg/dl in at least two separate measurements and (2) serum protein levels are normal (see Section V for serum protein adjustments).

A. Etiology

Spurious elevations of serum calcium are not uncommon, especially in cases in which venipuncture has been difficult; venous stasis may result in hemoconcentration and falsely elevate calcium concentration. In cases of mild hypercalcemia, the calcium determination should be repeated in blood samples obtained without the application of tourniquet. In addition, multiple serum determinations are essential. The mean level of all determinations should be used for diagnostic purposes and not the highest or lowest value(s) obtained.

Thiazide diuretics are frequently associated with mild increases in serum calcium. Calcium levels exceeding 12 mg/dl are rarely caused by thiazide use. Whether or not thiazide diuretics unmask an underlying parathyroid adenoma or increase calcium through some other mechanism remains controversial.

Malignancy is the most common cause of severe hypercalcemia (>12 mg/dl) in the elderly. Since hypercalcemia may represent the only early finding of malignancy in the elderly, a thorough evaluation of underlying malignancy is an essential part of the work-up. Lung and breast cancers are the most common tumors associated with hypercalcemia in the elderly.

Hyperparathyroidism, although a less common cause of hypercalcemia in the elderly, can result in hypercalcemia. In some cases the elevation in calcium occurs intermittently, making diagnosis difficult; repeating calcium determinations should help rule out spurious elevations in calcium.

Causes of Hypercalcemia in the Elderly

Common Causes

1. Thiazide diuretics
2. Malignancies
3. Hyperparathyroidism
4. Spurious

Uncommon Causes

1. Endocrine
   a. Hyperthyroidism
   b. Addison's disease
   c. Idiopathic elevation in postmenopausal women
2. Multiple myeloma
3. Paget's disease
4. Sarcoidosis
5. Chronic renal failure
6. High vitamin D intake
7. Milk alkali syndrome
8. Immobilization (?)

Hypercalcemia must not be attributed to immobilization without further evaluation. Although immobilization hypercalcemia has been described in chronically ill young men, it has only rarely been reported to occur in the elderly. Prolonged periods of immobilization, however, have been associated in the elderly with accelerated osteoporosis.

B. Clinical Findings

The symptoms and signs of hypercalcemia are listed.

Clinical Findings of Hypercalcemia

Reversible	Irreversible or Slowly Reversible
Lethargy, sleepiness	Conjuctival calcifications
Disorientation	Band keratopathy
Delirium, psychosis	Polyuria

(continued)

Clinical Findings of Hypercalcemia (*Continued*)

Reversible	Irreversible or Slowly Reversible
Weakness, weight loss	Renal calculi, nephrocalcinosis
Anorexia	
Nausea, vomiting	Osteopenia, bone cysts
Constipation	Pseudogout
Pruritus	Peptic ulcer
Bradycardia, ECG abnor-	Pancreatitis
malities (shortened Q–T	Skeletal muscle loss
interval)	Megacolon (rare)
Bone pain	
Azotemia	

Reversible findings are usually seen early in the course of illness; most are directly related to the degree of calcium elevation and usually are observed when calcium concentrations exceed 12.5 mg/dl. Findings are readily reversible on normalization of the serum calcium concentration.

The "irreversible or slowly reversible" findings are directly related to both duration and degree of calcium elevation. Although their presence usually indicates chronic hypercalcemia, they may even be found in patients with long-term mild elevations of serum calcium.

C. Diagnostic Approach

One methodical approach to the work-up of hypercalcemia is presented. Because of varying methods of determining parathyroid hormone (PTH), an elevated PTH level does not in itself make a diagnosis. Elevated levels may be associated with certain malignancies as well as renal insufficiency.

D. Treatment of Hypercalcemia

Acute treatment of hypercalcemia is not indicated unless the patient is symptomatic or calcium levels exceed 12 mg/dl.

Sodium diuresis is associated with increased calcium excretion and is a very effective short-term therapy for all causes of hyper-

Suggested Evaluation of Hypercalcemia in the Elderly

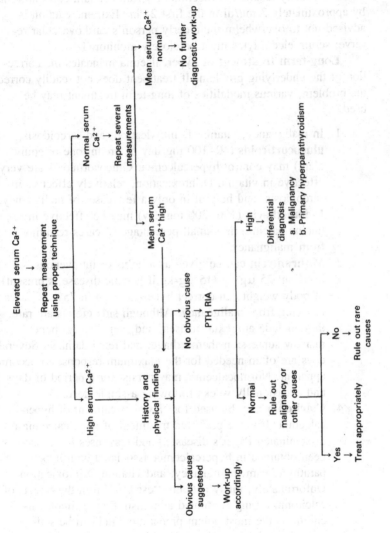

calcemia. Whenever permitted by the patient's cardiovascular status, 4–10 liters of normal saline should be infused intravenously in conjunction with 50–100 mg of furosemide every 2 hr. This regimen usually results in a decline of serum calcium concentration by approximately 3 mg/dl in the first 24 hr. Extreme caution is advised not to overwhelm the elderly person's cardiovascular reserve; serum electrolytes must be closely monitored.

Long-term treatment of hypercalcemia mandates the correction of the underlying problem. If treatment does not readily correct the problem, various modalities of long-term treatment may be used.

1. In malignancy, vitamin D intoxication, and sarcoidosis, **glucocorticoids** (40–100 mg/day of prednisone or equivalent) may control hypercalcemia. Glucocorticoids are very effective in vitamin D intoxication, relatively effective in sarcoidosis, and helpful in only a few cases of malignancy.

2. **Indomethacin** (100–200 mg/day) may be effective in reducing calcium in a small percentage of cases resulting from malignancy.

3. **Mithramycin** can be given as a bolus or intravenous infusion: 25 μg/kg (15 μg/kg if hepatic disease is present) of body weight can correct hypercalcemia in 75% of cases resulting from malignancy. Although side effects are rare, they include anorexia, nausea, and, very rarely, bone marrow supression, hemorrhage, and renal damage. Several days are often needed for the maximum response to become apparent. Normocalcemia may persist for a period of days and occasionally weeks following a single dose.

4. **Calcitonin** may be used for chronic treatment of hypercalcemia. It is the preferred treatment of hypercalcemia of disseminated Paget's disease; good responses have also been obtained in hypercalcemia associated with hyperparathyroidism, malignancy, and vitamin D intoxication. Unfortunately, many patients "escape" from the effects of calcitonin during continued administration. Salmon calcitonin is the most potent preparation and can be self-administered. This hormone is given subcutaneously on a daily schedule in doses not to exceed 8 MRC units/kg body weight; higher doses are not likely to have additional effect. Calcitonin may also be given either intravenously or intramuscularly every 6 hr for acute treatment of hypercalcemia.

5. **Estrogens** have been suggested for treating mild hypercalcemia in postmenopausal women. Recommended dosages are within physiological range.

V. HYPOCALCEMIA IN THE ELDERLY

Patients can be diagnosed as having hypocalcemia if (1) the
serum calcium concentration is less than 8.5 mg/dl in at least two
separate measurements and (2) serum protein levels are normal.

Since calcium is mostly bound to serum proteins, caution
should be exercised in interpreting serum calcium levels in the
presence of low levels of serum protein. It can be estimated that for
every 1 g/dl decline in serum protein levels, the serum calcium will
decline by 0.8 mg/dl. For accurate assessment of calcium status,
levels of ionized calcium can be obtained.

A. Etiology

The most common causes of hypocalcemia in the elderly are
listed.

Causes of Hypocalcemia in the Elderly

1. Vitamin D deficiency
2. Magnesium deficiency
3. Postablative hypoparathyroidism
4. Idiopathic hypoparathyroidism
5. Medullary carcinoma of the thyroid
6. Renal failure

Vitamin D deficiency in the elderly is usually the result of a
malabsorption syndrome. Long-term diphenylhydantoin use may
also result in a vitamin D deficiency because of altered hepatic
metabolism.

Magnesium deficiency is almost always associated with
alcoholism.

Postablative hypoparathyroidism may be the result of prior parathyroidectomy or thyroidectomy. Although this is usually observed within a few days after surgery, occasionally it becomes apparent only after a few months.

Idiopathic hypoparathyroidism is rarely seen in the elderly and usually is associated with the presence of other autoimmunoendocrinopathies (i.e., Hashimoto's thyroiditis, Addison's disease, pernicious anemia).

Medullary carcinoma of the thyroid usually is familial. Investigation of asymptomatic family members should be considered. Since it is associated with multiple endocrine adenomatosis syndrome (type II), a work-up for associated problems should be initiated.

B. Symptoms

Although mild hypocalcemia may be completely asymptomatic, psychological manifestations (i.e., irritability, mood changes, depression) and neurological symptoms (i.e., paresthesias, muscle cramps) may be noted. Symptoms of severe hypocalcemia include delirium, psychosis, tetany, and/or seizures. The Q–T interval on ECG may be prolonged.

On physical examination, findings of moderate and severe hypocalcemia include signs of hyperresponsiveness of the facial nerve such as twiching of facial muscles and contraction of muscles around the lip (**Chvostek's sign**) and spasm of the hand within 3 min after compression of the upper arm with a blood pressure cuff (**Trousseau's sign**).

Chronic disease may be associated with patchy hair, scaling skin, brittle fingernails, cataracts, candidiasis, osteopenia on bone X rays, and calcification of the basal ganglia.

C. Diagnostic Approach

One methodical approach to the diagnostic work-up of hypocalcemia is listed. Depending on historical information obtained, alternate approaches may be indicated. Caution is advised, however, to interpret laboratory results properly.

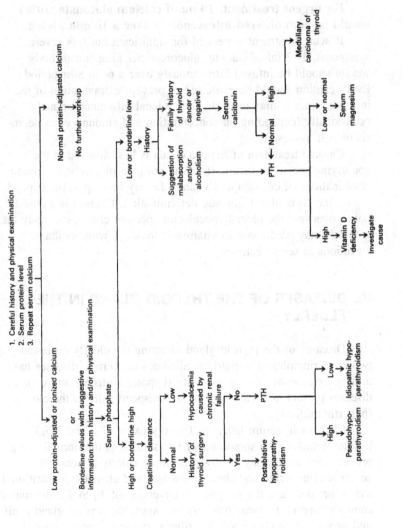

Evaluation of Hypocalcemia

1. Careful history and physical examination
2. Serum protein level
3. Repeat serum calcium

D. Treatment

Immediate treatment of hypocalcemia is not needed unless the patient is symptomatic or serum calcium levels decline to less than 7.5 mg/dl.

For **urgent treatment, 10 ml of calcium gluconate (10%) should be administered intravenously over a 10-min period.**

If acute treatment is needed for significant but less severe symptoms, 1–2 ml of calcium gluconate per kilogram of body weight should be infused intravenously over a 6- to 8-hr period. Every caution should be exercised to prevent extravasation of the infusate, since it may result in severe local inflammation and occasionally (depending on concentration and amount) even necrosis of soft tissues.

Chronic treatment of hypocalcemia is best directed at the underlying causative factor. If this is not possible, various dosage combinations of calcium and vitamin D may be employed, depending on the particular diagnostic determination. Caution is advised not to overtreat the older hypocalcemic person; changes in body fat with age may predispose to vitamin D toxicity, with resultant elevations in serum calcium.

VI. DISEASES OF THE THYROID GLAND IN THE ELDERLY

Diseases of the thyroid gland affecting the elderly represent two key principles of geriatric medicine; i.e., certain diseases have a higher prevalence during later life (hypothyroidism), and certain diseases present atypically and often nonspecifically in the elderly (hyperthyroidism).

Changes in serum levels of free thyroxine (T_4) and 3,5,3'-triiodothyronine (T_3) should never be ascribed to the normal aging process. Serum T_3 may be lower in many elderly because of accompanying acute and chronic diseases and changes in nutritional status that decrease the peripheral conversion of T_4 to T_3, the major source of serum T_3 formation. As one ages, the thyroid gland itself undergoes certain changes; e.g., fibrous tissue increases, colloid decreases, macro- and micronodules develop, and the thyroid becomes more retrosternal. As the thyroid gland begins to fail in its ability to maintain a euthyroid state, it begins to make more T_3. In fact, the ratio of $T_4 : T_3$ on circulating thyroglobulin decreases from

15:1 during youth to 5:1 later in life. The half-life of circulating T_4 increases from 5 days during childhood to over 9 days during the ninth decade of life. This has major clinical significance in replacing the hypothyroid elderly patient with thyroid hormone.

Common tests used to screen for thyroid function, total T_4 and T_3 resin uptake (T_3RU), both reflect T_4 levels and can be affected by many factors. Free hormone levels are the best indicator of thyroid function.

Factors Affecting T_3RU

Increase	Decrease
Androgen excess	Acute hepatitis
Glucocorticoids	Estrogens
Nephrotic syndrome	Oral contraceptives
Liver disease	Pregnancy
Acromegaly	Hypothroidism
Severe illness	Perphenazine
Stress	Acute intermittent por-
Diphenylhydantoin	phyria (hereditary X-
Hyperthyroidism	linked dominant trait)
Hereditary X-linked TBG	Chronic active liver disease
deficiency	Fluorouracil
Starvation	Marijuana abuse
Cirrhosis	Phenothiazines
Salicylates	Methadone
Anabolic steroids	Heroin
Aminoglutethimide	Clofibrate (Atromid-s®)
Chlorpropamide	
(Diabinese®)	
Cloflibrate (Atromid-s®)	
Sulfonamides	
Corticotropin	
Cortisone	
Methyltestosterone	
Nandrolone	
Norethandrolone	
Oxymethelone (Adroyd	
Anadrol-50®)	
Prednisone	
Testosterone	
Danazol	
L-Asparaginase	

Factors Affecting Total Serum Thyroxine Levels

Increase	Decrease	
Hyperthyroidism	Hypothyroidism	Cortisone
Elevated TBG	Decreased TBG	Iodides
Clofibrate	Diphenylhydantoin	Lithium
Estrogens	Heparin	Liothyronine
Methadone	Salicylates	sodium
Phenothiazines	Chlorpropamide	Methimazole
Progestins	Tolbutamide	Methylthiouracil
Thyroid	Exogenous T_3	Propylthiouracil
preparations	Aminosalicylic acid	Sulfonamides
Pregnancy	Ethionamide	Sulfonylureas
Iopanoic acid	Iodothiouracil	Dopamine
Amiodarone	6-Mercaptopurine	Phenylbutazone
Lopadate	Aminoglutethimide	Cholestyramine
	Nitroprusside	Cholestipol
	L-Asparaginase	Soybean flour

Although normal values of serum T_3 should not vary with age in healthy, well-fed persons, less elderly have values at the upper limit of the normal range. Whereas a serum T_3 of 180 mg/dl in a 30 year old is still within the variation of normal, this value may result in clinical symptoms in the 80 year old. As with all clinical practice, a total evaluation of the patient's clinical and laboratory status must be made; if questions still remain, further diagnostic testing should be pursued.

Thyrotropin-Releasing Hormone Stimulation Test

1. Helps distinguish among primary, pituitary, and hy-
   potholamic causes of hypothyroidism and can aid in
   diagnosing hyperthyroidism.
2. In the test, 400 μg TRH is given as a single i.v. bolus.
   In normal individuals the serum TSH concentration
   rises rapidly, reaching a peak in 20 to 30 min, and
   returns to basal values in 2 or 3 hr. In general, an
   increase of between 5 and 30 μU/ml TSH is accepted
   as normal.
3. In hypothyroidism of primary thyroid origin, the re-
   ponse to TRH is accentuated. A high basal TSH in a
   setting of clinical "hypothyroidism" makes the TRH

(continued)

Thyrotropin-Releasing Hormone Stimulation Test
(Continued)

test unnecessary. Pituitary causes for hypothyroidism are associated with a subnormal or absent response to TRH. Hypothalamic causes for hypothyroidism result in a normal but delayed response with peak concentrations not reached until 60 min after TRH administration.

4. Hyperthyroidism can be excluded by a normal TRH response. A flat response suggests hyperthyroidism; however, a subnormal response should not be taken as pathognomonic of clinically significant disease. Age, illness, and medications (phenytoin, thyroid hormone, iodide, glucocorticoids) can all blunt the TRH response.

A. Hyperthyroidism in the Elderly

The most common cause of hyperthyroidism in the elderly is a toxic nodular goiter. Graves' disease (diffuse toxic goiter) may present at any age but usually begins in the second, third, or fourth decade of life. Elderly men and women are affected equally, in contrast to the preponderance in women earlier in life.

1. Ten to 17% of thyrotoxic patients are 60 years of age or older.
2. Toxic nodular goiter is more common than Graves' disease in the elderly.
3. Fifty percent of thyrotoxic patients over age 60 have nodular goiters.
4. Twenty percent of elderly persons with hyperthyroidism have a normal-sized or nonpalpable thyroid gland.
5. Ophthalmologic signs are seen less frequently in the elderly.
6. Eighty percent of elderly thyrotoxic patients have cardiovascular abnormalities (palpitations, 60%; CHF, 66%; angina, 20%; atrial fibrillation, 40%).
7. Anorexia is common; weight loss and muscle wasting are usually profound.
8. Although diarrhea is rare, a correction of a preexisting constipation may be reported.
9. Tremor is often confused with "senile" tremor.
10. T_3 toxicosis caused by an autonomously functioning thy-

roid nodule occurs more commonly in the elderly. A serum T_3 determination is essential if clinical suspicion is high despite T_4 levels being within normal range.

The term "apathetic thyrotoxicosis" is used to describe the entity of hyperthyroidism, profound weight loss, apathetic facies, lethargy, depression, and lack of "classic" signs and symptoms of hyperthyroidism seen earlier in life. These patients often undergo extensive evaluation to exclude carcinoma, and cases wrongly treated for depression have been described. Profound weight loss and cardiovascular abnormalities are the usual presenting signs and should signal suspicion.

1. Evaluation of Hyperthyroidism in the Elderly

1. Comprehensive history and physical examination.
2. Laboratory testing: serum free T_4; serum T_3 by RIA if T_4 is normal; TRH test if suspicion of thyrotoxicosis remains despite equivocal laboratory testing.

2. Treatment of Hyperthyroidism

After diagnosis is confirmed, obtain a base-line complete blood count and differential and start on propylthiouricil (PTU) or methimazole (Tapazole®), titrating dosage to achieve euthyroidism. If there is an urgent need to induce euthyroidism promptly, oral iodides (Lugol's solution) may be prescribed; however, do this only after antithyroid medication has been given to block uptake of the iodide and thus decrease the potential for a worsening of thyrotoxicosis. Blockers of sympathetic action (i.e., propranolol) can also be useful adjuncts in highly symptomatic patients; side effects from these drugs must be carefully followed. Usually 100 mg of PTU (or 10 mg of Tapazole®) every 6 hr is effective; within 6 weeks most patients are euthyroid. The daily dose should be reduced to the lowest effective level (usually 50–300 mg PTU per day) while drug therapy is being given. Adverse reactions to these agents occur in fewer than 5% of patients and parallel the dose used. Agranulocytosis is the major adverse reaction, occurring in fewer than 0.5% of treated patients. Patients must be advised to have a CBC with differential any time a sore throat or fever develops. Since this occasionally fatal reaction can occur at any time, routine CBCs are not indicated.

Once euthyroidism is reached, arrangements for more definitive therapy with radioiodine (^{131}I) should be made. Some advocate

[131]I therapy before or even instead of giving initial antithyroid medications. Although this has certain merits, the usual weeks to months of delay in achieving euthyroidism following ^{131}I therapy and the risk, albeit small, of a radiation-induced throiditis with potential for thyroid storm limit this approach.

Surgery is rarely if ever indicated in the treatment of hyperthyroidism in the elderly.

B. Hypothyroidism in the Elderly

Many cases of hypothyroidism in the elderly escape recognition until late in the course of the illness. Estimates of the prevalence of hypothyroidism range from 1.0 to 10% depending on the population studied and interpretation of laboratory tests obtained.

Causes of Hypothyroidism in the Elderly

1. Autoimmune atrophy
2. Hashimoto's thyroiditis
3. Previous thyroid surgery
4. Previous radioactive iodine therapy
5. Recent discontinuation of thyroid hormone

Signs and Symptoms of Hypothyroidism Often Missed in the Elderly

Atrophic epidermis
Hyperkeratosis of stratum corneum
Coarse and thick hair
Alopecia
Loss of lateral third of eyebrows
Grooved nails
Congestive heart failure
Pericardial effusions
Alveolar hypoventilation
Occult pulmonary infections
Constipation
Cold intolerance
Lethargy
Altered mental status

---

1. Evaluation and Treatment of Hypothyroidism

1. Signs and symptoms of the disease are not affected by age; early recognition of these, however, is at times difficult for even the most skilled observer.

Multisystem Clinical Changes in Hypothyroidism

Skin

1. Atrophic epidermis
2. Hyperkeratosis of stratum corneum
3. Increased dermal mucopolysaccharides
4. Yellow skin secondary to poor hepatic conversion of carotene to vitamin A
5. Dermal vasoconstriction

Hair

1. Coarse and thick
2. Alopecia
3. Loss of lateral third of eyebrows

Nails

1. Brittle, thin, longitudinal grooves
2. Transverse grooves

Cardiovascular

1. Pericardial effusions
2. Congestive heart failure

Pulmonary

1. Alveolar hypoventilation and CO_2 narcosis
2. Decreased ventilation in response to hypoxia or hypercarbia
3. Occult pulmonary infections

Gastrointestinal

1. Decreased smooth muscle contractility resulting in constipation, impaction, dilation, obstruction, occult GI bleeding
2. Malabsorption (rare)
3. Ascites

---

2. A serum TSH determination provides the best indicator of thyroid hormone deficiency. An elevated serum TSH may

indicate, however, either a true hypothyroid state or a failing but presently compensated thyroid gland. In either case, replacement with thyroid hormone is indicated.

3. Thyroid hormone replacement must be given with caution, remembering that age prolongs the half-life of thyroxine. Twenty-five to 50 μg L–thyroxine is an appropriate starting dose; increments should be made only after a steady state has been reached (4–6 weeks in most cases).

4. When a daily dose of 75 μg L-thyroxine is reached, serum free T_4 and TSH levels should be obtained. Further increments should be made with close attention to the clinical state, as many elderly, especially those with compromised cardiovascular systems, will not tolerate replacement dosages into the range of "normal." Most elderly require between 100 and 125 μg p.o. L-thyroxine per day to maintain euthyroidism. A return of the serum TSH to "normal" indicates that metabolic balance has been achieved.

2. The "Failing Thyroid Gland"

Many elderly have a "failing but compensated" thyroid status. As the thyroid gland begins to have difficulty producing adequate amounts of thyroxine, TSH levels become elevated in an attempt to stimulate hormone production. In addition, more T_3 is produced by the thyroid gland, maintaining euthyroidism as long as possible. Although most agree that replacement therapy is indicated when serum TSH clearly rises into the range of hypothyroidism (reducing the risk of further nodule formation and missed hypothyroidism), the question of whether or not to treat those individuals with only slight elevations in serum TSH and normal serum T_4 and/or serum T_3 remains controversial.

VII. THYROID CANCER

Although the exact incidence of thyroid cancer is unknown, it is estimated that it occurs in 20–30 persons per million population. The actual prevalence of thyroid cancer cells at autopsy, however, is much higher. Fortunately, in most cases, cancers of the thyroid gland are relatively indolent and slow growing.

Classification of Thyroid Carcinoma
Well-Differentiated
1. Papillary
   a. Occult
   b. Intrathyroidal
   c. Extrathyroidal
2. Follicular

Undifferentiated
3. Medullary (extremely rare in the elderly)
4. Anaplastic

A relationship between cancer of the thyroid gland and prior radiation exposure earlier in life has been clearly established.

Papillary Thyroid Carcinoma
1. Most common cancer of the thyroid
2. Women affected three times more commonly than men
3. Can occur at any age, although peak incidence is between the third and fourth decades of life
4. Unencapsulated and slow growing
5. Commonly mixed with follicular elements within the tumor
6. Single lesion usual, although may occur in multiple foci
7. Tends to metastasize by lymphatic routes
8. Laminated calcific spheres (psammoma bodies) found in half of these tumors
9. Prognostic indicators include tumor size, direct local invasion, and distant metastases
10. Subdivided into three groups:
    a. Occult papillary carcinoma
       1. <1.5 cm in diameter
       2. Not palpable
       3. Despite cervical lymph node involvement, usually a benign disease
       4. Survival curves parallel those for age- and sex-matched normal subjects

---

(continued)

Papillary Thyroid Carcinoma (*Continued*)

    b. Intrathyroidal papillary carcinoma
       1. >1.5 cm in diameter but limited to the confines of the thyroid capsule
       2. 20% have bilateral thyroid involvement
       3. 50% exhibit local metastatic spread to lymph nodes
       4. Does not involve the trachea or other neck structures
       5. >85% 10-year and >70% 20-year survival
    c. Extrathyroidal papillary carcinoma
       1. Uncommon
       2. Extends beyond thyroid capsule into neck structures
       3. 50% 10-year and 40% 20-year survival noted
  11. Surgical removal indicated; the role of radioiodine therapy remains controversial; suppressive doses of thyroid hormone should be given if otherwise tolerated

Follicular Carcinoma

1. Accounts for 15% of all thyroid malignancies
2. Most common in women
3. Peak incidence in the fourth and fifth decades
4. Slow growing
5. Encapsulated
6. Range from well-differentiated, colloid-containing tumors to solid neoplasms with only slight follicular formation
7. Vascular and capsular invasion common
8. Bone, lung, and other soft tissue metastases most frequent
9. Once invasive, a 34% 10-year and 16% 20-year survival noted
10. Surgical removal of tumor, radioactive iodine therapy for treating local recurrences and distant metastases, and suppressive doses of thyroid hormone as tolerated are acceptable treatment modalities

Anaplastic Thyroid Cancer

1. Accounts for 5–10% of all thyroid carcinomas
2. Affects elderly persons most frequently
3. Women affected more commonly than men
4. Usually large and rapidly growing

(*continued*)

Anaplastic Thyroid Cancer (*Continued*)

5. Extends beyond the thyroid into adjacent structures
6. May coexist with papillary or follicular carcinoma, suggesting a malignant transformation of well-differentiated tumor
7. Presents with pain, tenderness, rapid growth, and local compression
8. Uniformly lethal with a median survival time of 6 to 8 months
9. Surgical removal of tumor mass and combination chemotherapy may reduce symptoms

---

VIII. DIABETES MELLITUS IN THE ELDERLY

A. Glucose Intolerance and Aging

Numerous studies have reported a decline in glucose tolerance with increasing age, usually first noted during the third decade of life. Following oral glucose ingestion, the 1-hr plasma glucose response has been shown to increase between 4 and 14 mg/dl (mean 9 mg/dl) per decade. The 2-hr plasma glucose value increases by 1 to 11 mg/dl (mean 5 mg/dl) per decade. The end result is a mean blood glucose value 1 hr after a 50-g oral glucose challenge of approximately 100 mg/dl for those aged 18 to 24 and 166 mg/dl for those aged 75–79.

Potential Complications of Increased Blood Glucose

Hyperosmolar states
Increased risk of atherosclerotic disease
Microvascular disease (neuropathy, nephropathy, and retinopathy)
Cataract formation
Infections
Premature death

Criteria for Diagnosing Diabetes in the Elderly

1. Fasting plasma glucose concentration must be equal to or greater than 140 mg/dl on more than one occasion

Or

2. Plasma glucose 2 hr following ingestion of 1.75 g/kg glucose (up to 75 g) must exceed 200 mg/dl and at least one other glucose value during the 2-hr oral glucose tolerance test must be greater than 200 mg/dl

*Based on National Diabetes Data Group (NDDG).

B. Managing the Elderly Diabetic

As previously discussed, it is often difficult to decide when treatment is necessary for the elderly person with glucose intolerance. Frequently, the diagnosis is made only after diabetes-related complications such as infections, neuropathy, and microvascular disease become problematic or the patient seeks medical care for a completely unrelated problem. In addition, controversy still exists as to whether untreated individuals with mild glucose intolerance will develop diabetes-related complications.

Almost half of elderly diabetic patients can be successfully managed with diet alone. Approximately 30% of elderly diabetics will require insulin some time in their course of illness; 20–25% can be treated with a combination of oral hypoglycemic agents and diet. Most clinicians will accept higher blood glucose levels when treating the elderly diabetic because of increased risks of hypoglycemia at this time of life. With the aging process, the ability to respond to stress and to maintain homeostasis becomes impaired. Elderly patients may not tolerate physiological stress as well as in their earlier life. Although many clinicians question whether "tight" control of blood glucose late in life retards complications that have already developed, it is well accepted that problems such as cataracts, neuropathy, and infections may be reduced with good blood glucose control.

Since the majority of elderly patients have type II diabetes, management should be individualized to meet the patient's life style, nutrition, and health care needs. Although insulin must be present to insure normal glucose, lipid, and protein metabolism, endogenous insulin may suffice if a dietary regimen can be adhered to. In addition, increased activity as tolerated can improve glucose utilization with a resultant lowering of the blood glucose level. A careful medical evaluation is essential to establish exercise limits in light of increased interacting cardiologic and neurological problems at this time of life. Diet alone should be tried initially in all elderly patients who are not hyperosmolar or suffering from excessive glycosuria and do not have blood glucose levels in excess of 300 mg/dl. Diets should be relatively low in calories with approximately 60–70% in the form of carbohydrate, 15–20% protein, and 20–25% fat. The use of supplemental fiber has been shown to improve glucose tolerance in certain cases by evening out glucose absorption.

Although insulin therapy may not initially be recommended for the obese individual, age is not a contraindication for insulin use. Consideration must be given to life style, ability to administer accurately a prescribed dose, dietary compliance (missed meals increase the risk of hypoglycemia), and ability to monitor glucose control. Split regimens utilizing NPH and regular insulin have the advantage of providing more uniform glucose control. Two daily injections of insulin, however, may not be tolerated by all elderly patients. In general, insulin should be started in low dosages. This is particularly important as many elderly have labile levels of blood glucose, changing rapidly with acute illness, changes in diet, and activity.

Insulin Preparations

Action	Type of insulin	Peak activity (hr)
Rapid	Regular crystalline	2–4
	Humulin R	2–4
	Actrapid	2.5–5
	Semitard (Semilente)	5–10
Intermediate	NPH	6–12
	Humulin N	6–12
	Lente	7–15
	Monotard (Lente)	7–15
Long-acting	Ultralente	18–24

1. Guidelines for Managing the Elderly Diabetic

1. Blood glucose levels should be kept in a range where symptoms (i.e., polyuria, polydipsia, weight loss, vision changes) are controlled. Blood glucose levels usually should not exceed 220 mg/dl.
2. Avoid too "tight" control; elderly persons are particularly prone to complications from hypoglycemia.
3. Diet therapy alone should be tried initially in all elderly patients who are not hyperosmolar or suffering from excessive glycosuria and do not have blood glucose levels in excess of 300 mg/dl. Diets should be relatively low in calories, with approximately 60–70% in the form of carbohydrate, 15–20% protein, and 20% fat. Supplemental fiber may improve glucose tolerance in certain cases by evening out glucose absorption.
4. Physical activity should be encouraged as tolerated, as this helps reduce insulin requirements.
5. **Insulin.** Before starting insulin therapy, the patient's life style, ability to administer a prescribed dose, dietary compliance (missed meals increase risk of hypoglycemia), and ability to monitor glucose control should be considered. Split regimens utilizing NPH and regular insulin have the advantage of providing more uniform control. By totaling the daily requirement of regular, short-acting insulin, a starting dose can usually be determined based on two-thirds of the dose being given in the morning and one-third given in the afternoon. The morning dose should consist of two-thirds NPH and one-third regular insulin and the afternoon dose of half NPH and half regular insulin. Dosage of insulin must be adjusted as necessary. Although this regimen is usually successful, the two daily injections of insulin may not be tolerated by all elderly patients. In general, insulin should be started in low dosages; many elderly have labile levels of blood glucose, changing rapidly with acute illness, changes in diet, and activity.
6. **Oral hypoglycemic agents.** These are particularly effective when fasting blood glucose levels do not exceed 220 mg/dl. Choice of agent should be based on individual characteristics. Second-generation agents have many distinct characteristics that are an advantage when treating the elderly diabetic patient.

7. All elderly diabetics must be carefully screened for age and diabetes-associated illness, including hyperlipidemias, renal insufficiency, and hypothyroidism.
8. A yearly ophthalmologic evaluation is recommended as a minimum.
9. Podiatric treatment is frequently required to minimize risk of complications.

2. Oral Hypoglycemic Agents

Oral hypoglycemic agents are often effective in achieving blood glucose control, particularly when the fasting blood sugar is less than 220 mg/dl. Although numerous preparations are available, individual characteristics must be considered. Tolazamide has the advantage of having a short half-life, minimizing the risk of hypoglycemia associated with longer-acting sulfonylureas. In addition, its mild diuretic effect is often helpful in the elderly. A high drug–drug interaction is an undesirable side-effect, however. Long-acting preparations such as chlorpropamide increase the risk of prolonged hypoglycemia if an excessive dosage in taken or if a meal is missed. The latter also has the potential for inducing fluid retention, thereby limiting its usefulness. Since tolbutamide has a short half-life and often requires two or three daily doses, compliance is a major impediment.

Second-generation agents have distinct characteristics that are an advantage when treating elderly diabetic patients. Glipizide is particularly useful because it is metabolized by the liver into inactive metabolites, eliminating problems resulting from age-related changes in renal function. Hypoglycemia and drug–drug interactions are rare. Recent data also suggest that glipizide induces a physiologic response to meals. Insulin is elevated following meals with a return to baseline during fasting. This unique characteristic further reduces the chance of developing hypoglycemia. Low cost is an additional advantage of second-generation agents. These agents are usually successful when given once daily, further improving compliance. Many patients who fail on first-generation oral agents may be successfully treated with these new medications. (See table on p. 171.)

It is essential that patients be carefully screened for age- and diabetes-associated illness, including hyperlipidemias, renal insufficiency, and hypothyroidism. An ophthalmologic evaluation should be done yearly or more often as indicated. Foot care cannot be

Oral Agent Characteristics

	Duration of action	Recommended starting dosage	Dosage range (mg/day)	Tablet size (mg)	Comments
Acetohexamide (Dymelor®)	12–24 hr	250 mg once daily	25–1500	250, 500	Short duration of action; relatively inexpensive; mild diuretic effect
Chlorpropamide (Diabinese®)	Several days	100 mg once daily	100–500	200, 250	Long duration of action; fluid retention; renal excretion
Tolazamide (Tolinase®)	12–24 hr	100 mg once daily	100–750	100, 250, 500	Mild diuretic effect
Tolbutamide (Orinase®)	6–24 hr	250–500 mg once daily or in divided dosage	250–3000	250, 500	Short duration of action
Glyburide (Micronase® or Diabeta®)	24 hr	2.5–5 mg once daily in most cases	2.5–20	1.25, 2.5, 5.0	Second generation; metabolites excreted equally in urine and feces; mild diuretic effect; low cost; Drug–drug interactions rare
Glipizide (Glucotrol®)	24 hr	2.5 mg once daily in most cases	2.5–40	5, 10	Second generation; drug–drug interactions rare; mild diuretic effect; low cost; hypoglycemia rare; hepatic metabolism to inactive metabolites

overemphasized, especially in patients with neuropathy and/or peripheral vascular disease. Podiatric referral is frequently required.

3. Monitoring Treatment in the Elderly

The best indicator of satisfactory treatment is a normal blood glucose level. Although home glucose monitoring is becoming more common, many elderly cannot use or afford the instrumentation involved in this procedure. Urine tests provide limited information with a time lag of several hours. A general assessment of control can, however, be gained by checking urines for glucose several times a day as long as there has been prior correlation of urine determinations with blood glucose levels. Caution is advised as many elderly have difficulty in reading "colorimetric" scales because of age-related changes in color vision.

Frequent visits to the physician are advised when starting the elderly patient on treatment. In general, fasting blood glucose levels should range between 120 and 140 mg/dl with postprandial blood glucose levels between 160 and 220 mg/dl. This range affords protection against hypoglycemia and minimizes the risk of complications resulting from hyperglycemia.

Adjustments in treatment may be required during times of physiological stress. Patients unable to eat must be advised to reduce the dosage of insulin or oral hypoglycemic agent and to attempt dietary supplementation as much as possible. Dehydration and glycosuria must always be avoided.

Although measurements of hemoglobin A_{1c} (glycosylated hemoglobin) are useful in monitoring blood glucose control in young diabetics, this parameter is not easily interpretable in the elderly. Recent studies have reported many elderly to have elevated HbA_{1c} levels despite the absence of glucose intolerance. Factors other than plasma glucose are capable of altering HbA_{1c} levels.

Factors Capable of Altering HbA_{1c}

Reduced
Hemolytic anemia
Chronic blood loss
Presence of abnormal hemoglobins (S,C,D)

(continued)

Elevated

Elevated hemoglobin F
Hyperglycemia
Thalassemia
Chronic renal failure
Dialysis
Splenectomy
Elevated triglycerides

In addition, other factors can affect HbA_{1c} levels.

IX. HYPERLIPIDEMIA AND THE ELDERLY

There are five primary and several secondary forms of hyperlipidemia.

Type	I	II a	II b	III	IV	V
Elevated level of	Chylomicrons	LDL (B)	LDL (B) and VLDL (pre-B)	Abnormal LDL	VLDL (pre-B)	Chylomicrons and VLDL
Other description	Exogenous (fat-induced) hyperlipemia	Essential hypercholesterolemia		Broad B disease	Endogenous (carbohydrate induced) hyperlipemia	Mixed hyperlipemia
Prevalence	Rare	Relatively common		Relatively uncommon	Most common	Uncommon
Appearance of plasma after overnight refrigeration	Creamy layer over infranate on standing	Clear	Clear or turbid	Turbid ± cream layer	Turbid, lactescent	Creamy layer over milky infranate on standing
Cholesterol elevation	+	+++	+++	++	0-+	+
Triglyceride elevation	+++	0	+	+	++	+++

Clinical Features of Hyperlipidemias

Type	Clinical Findings
I	Eruptive xanthomas; lipemia retinalis; hepatosplenomegaly; pancreatitis; abdominal pain

(*continued*)

Clinical Features of Hyperlipidemias (*Continued*)

II	Xanthelasma; extensor tendon xanthomas (hands, elbows, Achilles tendon, tibial tuberosity); tuberous and planar xanthomas; premature corneal arcus; premature atherosclerosis
III	Orange–yellow planar xanthomas in palmar creases; tubereruptive xanthomas; premature corneal arcus; premature atherosclerosis on coronary and peripheral arteries
IV	Eruptive xanthomas; premature atherosclerosis; obesity; hepatosplenomegaly; abnormal glucose tolerance; occasionally, abdominal pain
V	Abdominal pain; eruptive xanthomas

Secondary Causes of Hyperlipidemia

1. Diabetes
2. Hypothyroidism
3. Nephrotic syndrome
4. Dysproteinemia (multiple myeloma, macroglobulinemia, lupus erythematosus)
5. Liver disease (obstructive)
6. Ethanolism
7. Drugs
8. Pregnancy and exogenous estrogens
9. Obesity
10. Cholesterol-rich diet (saturated fats)
11. Acute intermittent porphyria
12. Uremia

Medications Capable of Affecting Lipid Levels

	Cholesterol	Triglycerides
1. Estrogens	—	↑
2. Glucocorticoids	—	↑
3. β blockers	—	↑ ↑
4. Thiazide diuretics	↑	↑ ↑

Normal Total Cholesterol Values (mg/dl)*

White males				White females (non-sex-hormone users)				
Age		Percentiles			Age		Percentiles		
(years)	Mean	5	50	95	(years)	Mean	5	50	95
0–4	155	114	151	203	0–4	156	112	156	200
5–9	160	121	159	203	5–9	164	126	163	205
10–14	158	119	155	202	10–14	160	124	158	201
15–19	150	113	146	197	15–19	157	120	154	200
20–24	167	124	165	218	20–24	164	122	160	216
25–29	182	133	178	244	25–29	171	128	168	222
30–34	192	138	190	254	30–34	175	130	172	231
35–39	201	146	197	270	35–39	184	140	182	242
40–44	207	151	203	268	40–44	194	147	191	252
45–49	212	158	210	276	45–49	203	152	199	265
50–54	213	158	210	277	50–54	218	162	215	285
55–59	214	156	212	276	55–59	231	173	228	300
60–64	213	159	210	276	60–64	231	172	228	297
65–69	213	158	210	274	65–69	233	171	229	303
70+	207	151	205	270	70+	228	169	226	289

*Modified from U.S. Dept. of Health and Human Services.

Normal LDL-Cholesterol Values (mg/dl)*

White males				White females (non-sex-hormone users)				
Age		Percentiles			Age		Percentiles		
(years)	Mean	5	50	95	(years)	Mean	5	50	95
0–4	—	—	—	—	0–4	—	—	—	—
5–9	93	63	90	129	5–9	100	68	98	140
10–14	97	64	94	132	10–14	97	68	94	136
15–19	94	62	93	130	15–19	95	60	93	135

(continued)

Normal LDL-Cholesterol Values (mg/dl)* (Continued)

Age (years)	White males				Age (years)	White females (non-sex-hormone users)			
	Mean	Percentiles				Mean	Percentiles		
		5	50	95			5	50	95
20–24	103	66	101	147	20–24	98	62†	98	136‡
25–29	117	70	116	165	25–29	106	70	103	151
30–34	126	78	124	185	30–34	109	67	108	150
35–39	133	81	131	189	35–39	119	76	116	172
40–44	136	87	135	186	40–44	125	77	120	174
45–49	144	98	141	202	45–49	130	80	127	187
50–54	142	89	143	197	50–54	146	90	141	215
55–59	146	88	145	203	55–59	152	95	148	213
60–64	146	83	143	210	60–64	156	100	151	234
65–69	150	98	146	210	65–69	162	97	156	223
70+	143	88	142	186	70+	149	96	146	207

*Modified from U.S. Dept. of Health and Human Services.
†Tenth percentile.
‡Ninetieth percentile.

Normal Total Triglyceride Values (mg/dl)*

Age (years)	White males				Age (years)	White females (non-sex-hormone users)			
	Mean	Percentiles				Mean	Percentiles		
		5	50	95			5	50	95
0–4	56	29	51	99	0–4	64	34	59	112
5–9	56	30	51	101	5–9	60	32	55	105
10–14	66	32	59	125	10–14	75	37	70	131
15–19	78	37	69	148	15–19	72	39	66	124
20–24	100	44	86	201	20–24	72	36	64	131
25–29	116	46	95	249	25–29	75	37	65	145
30–34	128	50	104	266	30–34	79	39	69	151
35–39	145	54	113	321	35–39	86	40	73	176
40–44	151	55	122	320	40–44	98	45	82	191
45–49	152	58	124	327	45–49	105	46	87	214
50–54	152	58	124	320	50–54	115	52	97	233
55–59	141	58	119	286	55–59	125	55	106	262
60–64	142	58	119	291	60–64	127	56	105	239
65–69	137	57	113	267	65–69	131	60	112	243
70+	130	58	111	258	70+	132	60	111	237

*Modified from U.S. Dept. of Health and Human Services.

A. High-Density Lipoprotein

High-density lipoprotein (HDL) is thought to remove cholesterol from extrahepatic tissues as well as to compete with LDL-cholesterol binding. An inverse relationship between levels of HDL-cholesterol and the incidence of coronary artery disease has been known since the early 1950s.

High-density lipoprotein has been further divided into HDL_2 and HDL_3 subfractions. It is the HDL_2 concentration that best correlates inversely with coronary artery disease. In patients recovering from a myocardial infarction, HDL_2 levels are reportedly decreased. Women have higher HDL_2 levels than men; regular exercise also increases HDL_2 levels.

In men, levels of HDL-cholesterol decrease after the first decade of life until the seventh decade, after which slight increases may be noted. In women, increases in HDL-cholesterol have been noted between the third and seventh decades of life.

Factors Capable of Altering HDL Levels

Increase HDL	Decrease HDL
1. Estrogen	1. Androgens
2. Prazosin	2. β blockers
3. Phenytoin	3. Diazepam
4. Phenobarbital	4. Probucol
5. Nicotinic acid	5. Diabetes mellitus
6. Exercise	(poorly controlled)
7. Insulin	6. Myocardial infarction
8. Weight loss	7. Upper respiratory
9. Alcohol	infection
10. Terbutaline	8. Postoperatively
	(cholecystectomy)
	9. Cigarette smoking
	10. Obesity

B. Evaluation of Hyperlipidemia

A fasting lipid profile, including levels of HDL-cholesterol, is necessary for accurate diagnosis. An overnight "standing plasma test" can be helpful in making a diagnosis prior to laboratory confirmation.

Normal HDL Cholesterol Values (mg/dl) for White Males and Females*

Age (years)	Males				Females (non-sex-hormone users)			
		Percentiles				Percentiles		
	Mean	5	50	95	Mean	5	50	95
5–9	55	38	54	74	53	38	54	74
10–14	55	37	55	74	52	37	52	70
15–19	46	30	46	63	52	35	51	73
20–24	45	30	45	63	52	37†	50	68‡
25–29	45	31	44	63	56	37	55	81
30–34	45	28	45	63	55	38	55	75
35–39	43	29	43	62	55	34	52	82
40–44	44	27	40	67	57	33	55	87
45–49	45	30	45	64	58	33	56	86
50–54	44	28	44	63	60	37	59	89
55–59	48	28	46	71	59	36	58	86
60–64	51	30	49	74	62	36	60	91
65–69	51	30	49	78	61	34	60	89
70+	51	31	48	75	60	33	60	91

*Modified from U.S. Dept. of Health and Human Services.
†Tenth percentile.
‡Ninetieth percentile.

Hypercholesterolemia caused solely by elevated levels of HDL-cholesterol (extremely rare) would not warrant any treatment. On the other hand, hypercholesterolemia resulting from increased LDL-cholesterol levels with decreased HDL-cholesterol concentrations warrants aggressive therapy.

Total cholesterol and triglyceride levels should be obtained while fasting, 10–15 hr after the last meal. Hyperlipidemia is generally defined by values in the upper 5% of the Lipid Research Clinic's tables.

Although LDL-cholesterol is a better predictor than total cholesterol for atherogenesis, its direct measurement is difficult and requires ultracentrifugation.

An estimate of LDL-cholesterol can be calculated using the following formula:

$$\text{LDL-cholesterol} = \text{serum cholesterol} - \frac{\text{serum triglyceride}}{5} - \text{HDL-cholesterol}$$

A specific diagnosis can usually be made by history and physical findings, the appearance of an overnight refrigerated plasma specimen, and levels of specific lipids.

A lipid electrophoresis does not usually provide additional information.

Secondary causes of hyperlipidemia must always be ruled out. Rarely, patients will need to be referred to specialized centers for evaluation of rare lipid disorders.

C. Treatment

Although there is clearly an increased risk of atherogenesis in populations with hypercholesterolemia, it remains unclear what level requires treatment. Although treatment must be carefully considered regardless of one's age, possible benefits must be weighed against risks of treatment. The Lipid Research Clinics Prevalence Study, done in 1979, of 5000 U.S. and Canadian men and women reported 95th-percentile levels of cholesterol of 275 mg/dl in older men and 300 mg/dl in older women. The 95th-percentile levels for triglyceride ranged from 200–300 mg/dl in men and 130–230 mg/dl in women.

Despite these limits, little is known in the elderly population regarding the benefit of treating borderline elevations in cholesterol and/or triglyceride levels.

D. Methods of Treatment

1. Diet

Although it is difficult to decrease serum lipids by diet alone, this form of therapy should be encouraged. The following represent some guidelines for diet therapy.

1. Restrict calories sufficiently to obtain weight reduction, with the goal being average body weight for age.
2. Dietary fat should be reduced to 15–20% of total calories.
3. Dietary carbohydrates should be increased to 65–70% of total calories.
4. Dietary protein should be limited to 25% of total calories (emphasizing vegetable protein).
5. Dietary cholesterol should be restricted to 300 mg/day and

if necessary to 100 to 200 mg/day (an average diet contains approximately 500 mg).
6. Polyunsaturated fats should be substituted for saturated fats with a ratio of saturated/polyunsaturated fats less than 1.0. If possible, a ratio as low as 0.3 should be encouraged.
7. Reduce alcohol intake, primarily in patients with elevations of triglycerides.
8. Increased dietary fiber may be useful in lowering plasma lipid levels.

2. Medications

a. Cholestyramine. Cholestyramine is an insoluble and non-absorbable anion-exchange resin that binds bile. Since bile is not reabsorbed, additional cholesterol is metabolized and excreted. A fall in LDL-cholesterol concentration usually occurs within 1 to 2 weeks. Although LDL-cholesterol concentrations usually decrease 15–35%, occasional elevations in triglyceride levels may be seen. Side effects of particular concern in the aged include constipation, nausea, bloating, indigestion, small bowel obstruction, and malabsorption of drugs including fat-soluble vitamins, anticoagulants, thyroxine, certain antibiotics, digitalis, phenobarbital, phenylbutazone, and thiazide diuretics.

b. Clofibrate (Atromid-S®) and Gemfibrozil (Lopid®). Clofibrate and gemfibrozil are quite similar and appear to alter lipid levels by affecting lipolysis, reducing the supply of free fatty acids going to the liver. Although these medications can lower both cholesterol and triglyceride levels, their predominant influence is on triglycerides. Clofibrate is excreted mainly by the kidneys and has a half-life of about 12 hr in the presence of normal renal function; clofibrate should be monitered carefully in the elderly who have reduced renal clearance. In cases of significant renal insufficiency, clofibrate may need to be withheld. Although studies from the World Health Organization suggest an increase in mortality from nonischemic heart disease following clofibrate use, the medication did reduce serum cholesterol levels with an associated decrease in the incidence of nonfatal myocardial infarction. Adverse effects include a rise in biliary cholesterol and an increased risk of cholelithiasis and cholecystitis. Other side effects include headaches, nausea, diarrhea, and myositis (rare increase in CPK).

c. Probucol. Probucol increases fecal excretion of bile acids and is thought to lower LDL and HDL lipoproteins by inhibiting cholesterol synthesis. Approximately 50–70% of patients will improve on probucol with a 15–35% reduction in plasma cholesterol levels.

The drug appears to have little toxicity and is generally well tolerated. The most common and usually transient side effects include nausea, pyrosis, abdominal pain, loose stool, headaches, palpitations, and syncope.

d. Nicotinic Acid. Nicotinic acid is a vitamin that decreases secretion of VLDL. This leads to a fall in LDL-cholesterol and a rise in levels of HDL-cholesterol. Numerous side effects result in poor patient compliance. Although gastric distress, cutaneous flushing, and pruritis are frequently seen, they are often transient. Other side effects include nausea, vomiting, diarrhea, dry skin, and dizziness. The patient should be counseled regarding these side effects. Taking the medication with meals may reduce some of these problems; aspirin may be used prophylactically to prevent flushing. Hyperglycemia and hyperuricemia may be exaggerated by nicotinic acid. Nicotinic acid has been shown to lower triglyceride levels by 35–60% and cholesterol by 30%.

e. D-Thyroxine. D-Thyroxine and other synthetic analogues have been developed in the attempt to isolate the effect of cholesterol reduction from the hypermetabolic effects of L-thyroxine. D-Thyroxine is not devoid of these effects and acts primarily by increasing the conversion of cholesterol to bile salts in the liver as well as by increasing LDL receptors. The adverse effects are those of thyroid hormone. Although some advocate concomitant use of propranolol, side effects of the latter are numerous in the elderly and often outweigh the benefit. Elevation of liver enzymes and potentiation of coumarin anticoagulation may also occur. In general, D-thyroxine is not a good choice for lipid control in the elderly.

f. β-Sitosterol. Although β-sitosterol is structurally related to cholesterol, it is poorly absorbed from the intestine and impairs absorption of cholesterol from the gut. Reductions in cholesterol are not universal and generally do not exceed 20%. In addition, a significant drop in HDL-cholesterol may occur. Constipation is a rare side effect. This drug is contraindicated in β-sitosterolemia, a rare condition resulting from the abnormal absorption of plant sterols.

g. Norethindrone and Oxandrolone. Norethindrone and oxandrolone are second-line drugs used in severe hyper-triglyceridemia unresponsive to diet and treatment with first-line antihyperlipidemic agents. Both have androgenic properties and may result in fluid retention, hypertension, and hirsutism. They are thought to increase hepatic lipase activity and thereby increase disposal of triglycerides.

h. Mevinolin. Mevinolin is an experimental drug that acts as a competitive inhibitor of 3-hydroxy-3-methylglutarylcoenzyme A (HMG-CoA) reductase. Mevinolin is thought to increase receptor-mediated LDL catabolism. An approximately 20–30% reduction in total plasma cholesterol has been reported. Side effects may include abdominal pain, diarrhea, nausea, headaches, insomnia, rash, and fatigue.

i. Combined-Drug Regimens. In cases in which single-drug therapy proves unsuccessful, combination therapy may be indicated. The risk–benefit ratio, however, must be carefully con-sidered. Resin binders used in combination with a HMG-CoA reductase inhibitor may have a synergistic effect. Other drugs such as clofibrate and nicotinic acid used with resin binders may also be more effective than single-agent therapy. The use of such regimens should be monitored carefully because of a possible higher inci-dence of adverse effects. (See table on p. 183.)

X. SEXUAL DYSFUNCTION IN ELDERLY MEN

Approximately 25% of men at age 65 and 55% at age 75 suffer from impotence. In clinical settings, incidence varies accord-ing to whether or not a sex history is obtained or whether one relies on patient-initiated reports.

There is also a difference of opinion in the length of time recommended for a couple to establish an adequate sexual adjust-ment before seeking help; steady relationships often result in an appreciable spontaneous recovery rate.

Lipid-Lowering Medications and Their Effects*

Medications	Dosage	Site of action	Primarily lowers	VLDL	LDL	HDL
First line						
Cholestyramine (Questran®)	12–24 g/day	Intestine	Cholesterol	↕	→	↕
Colestipol (Colestid®)		Intestine	Cholesterol	↕	→	↕
Clofibrate (Atromid-S®)	2 g/day	Systemic	Triglycerides	→	→	→
Gemfibrozil (Lopid®)	1.2 g/day	Systemic	Triglycerides	→	→	←
Probucol (Lorelco®)	1 g/day	Intestine and systemic	Triglyceride	↕	→	→
Nicotinic acid (niacin)	1–4 g/day	Systemic	Triglyceride Cholesterol	→	→	→
Second line						
D-Thyroxine (Choloxin®)	2–8 mg/day	Systemic	Cholesterol Triglycerides	↕	→↑←	↓↑→
β-Sitosterol	3–8 g/day	Intestine	Cholesterol	→	←	→
Neomycin	1.5–2 g/day	Intestine	Cholesterol	→		
Norethindrone (Norlutate®)	5 mg/day	Systemic	Triglycerides	→	←	
Oxandrolone (Anavar®)	7.5 mg/day	Systemic	Triglycerides	→		→
Investigational						
Mevinoline	40–80 mg/day	Systemic	Cholesterol	↕	→	↕

*Adapted from *Contemporary Geriatric Medicine*, volume II, New York, Plenum Press, Gambert, S. (ed), 1986.
†↑, increase; ↓, decrease; ↔, no change.

A. Etiology

Independent of whether organic etiologic factors are found, psychological factors can be detected in almost every impotent male. Depression, anxiety, and anger can interfere with the sexual response cycle.

A common problem observed in the aging male is the "widower's syndrome." This is commonly reported in older men who have not had coitus for a period of time following separation from a steady partner by either death or divorce.

Despite the opportunity for sexual activity and an interest to perform satisfactorily, there may not be sufficient time to redevelop the levels of integrated organismic functioning necessary for full penile erection. The phenomenon is analogous to that of any natural function following periods of marginal activity.

The most common organic cause of impotence is drug use. A partial list of drugs proven to cause, or strongly suspected of causing, this problem is presented below.

Some Commonly Used Drugs Associated with the Development of Impotence

Drug group	Agent
Psychotropic	Phenothiazines, butyrophenones, barbiturates, benzodiazepines, phenytoin, carbamazepine, narcotic analgesics, tricyclic antidepressants, lithium carbonate, cannabis,* alcohol (abuse)
Antihypertensives	Guanethidine, reserpine, methyldopa, clonidine,* β-adrenegic blockers,* prazosin*
Other	Atropine, benzatropine, phenoxybenzamine, clofibrate, metoclopramide, estrogens, spironolactone, adrenal steroids (high doses), cimetidine, serotonin antagonists*

*Suspected but not adequately proven association.

Several diseases associated with impotence in elderly men are listed.

Diseases Associated with Impotence

System	Disease
Endocrine–metabolic	Hypogonadism, hypo- and hyperthyroidism, increased endogenous estrogens (tumors), Addison's disease, pituitary adenomas, diabetes mellitus, hemochromatosis
Genital	Castration, radical prostatectomy, Peyronie's disease, phimosis
Neurological	Brain lesions (temporal lobe), spinal cord lesions, pelvic nerve lesions, limbic system lesions
Vascular	Sickle cell anemia, Leriche syndrome
Mixed, chronic	Renal failure, hepatic cirrhosis, malignancies, chronic infections

---

B. Evaluation of Impotence in the Elderly Man

Most experts recommend a thorough evaluation of patient complaints to include not just the loss of sexual interest, function, or frequency but also somatic and emotional equivalents that might contribute to the development of sexual problems.

Initial Evaluation of Sexual Problems in the Elderly

1. History
    a. Loss of sexual interest, potency, etc.
    b. Symptoms of depression
    c. Symptoms of anxiety

(continued)

Initial Evaluation of Sexual Problems in the Elderly
(*Continued*)

    d. Physical symptoms (skeletal pains, UTI symptoms, etc.)

    e. Information on drug usage

2. Physical examination, including genital examination
3. Sexual history
4. Counseling or referral to specialist as indicated
5. Depending on the results of 1–3, the patient should be evaluated by one or more of the following specialists:

    a. Psychiatrist or psychologist

    b. Internist or endocrinologist

    c. Urologist

History of medication use should always be obtained. A physical examination, including a genital examination, should be performed by the physician.

It is essential that all elderly men who experience difficulty in sexual performance and who are concerned enough to seek professional care undergo a careful diagnostic evaluation. Symptoms must not be automatically attributed to the process of aging, nor should treatments be initiated prior to careful investigation. If necessary, the patient should be referred to the appropriate specialist. A decline in serum testosterone levels has been associated with illness. Whether or not "normal" aging affects free testosterone levels remains controversial, although all agree that levels remain within the range of normal as reported by most laboratories.

C. Treatment

The treatment of impotence depends on the probable cause and must be individualized. If a psychological cause is strongly suspected, or if an organic cause is not apparent (or strongly suspected), the following scheme referred to as the PLISSIT model may be utilized. It represents four graduated steps or alternatives that might be applied to the patient's sexual concerns: permission, limited information, specific suggestions, and intensive work-up and/or therapy. Permission granted by the clinician may be all the patient needs to reverse an incipient sexual dysfunction. Sometimes the approval of an authority figure is all that is needed to resume

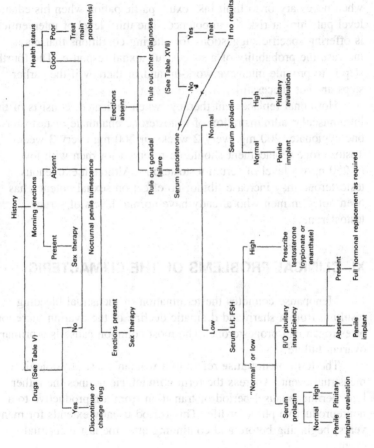

Evaluation of Impotence in the Elderly Man

sexual activity or to remotivate the patient's sexual desire and can even restore full penile tumescence. In other cases, a second step is needed, i.e., the giving of limited information. In many cases, the older adult needs to be informed that sex is a coexistent with life itself and not just a function of the young, employed, or healthy. Information given regarding ethanol's effect on sexual interest and motivation can be used by the patient either to restore potency when necessary or to limit his sexual participation when his ethanol level puts him at risk for impotence. The third level of intervention is offering specific suggestions for altering conditions that may increase the probability of a successful sexual response. The fourth step is to provide intensive work-up and/or therapy if the earlier steps are not successful.

Hormonal replacement therapy, when indicated, consists of the intramuscular administration of testosterone enanthate or testosterone cypionate, 200 mg every 2 weeks or 300 mg every 3 weeks. Testosterone replacement should be reserved for men with low (<250 ng/dl) level of serum testosterone. Although exogenous testosterone may increase libido, no effect on sexual potency has been noted in men who already have normal levels of serum testosterone.

XI. CLINICAL PROBLEMS OF THE CLIMACTERIC

Menopause connotes the termination of menstrual bleeding resulting from a sharp and dramatic decline of the ovarian secretion of estrogen and progesterone. The most common cause is a primary ovarian failure.

The term **menopause** refers to a woman's last physiological menstrual event, whereas the term **climacteric** defines the rather long perimenopausal period of transition from a reproductive to a nonreproductive phase of life. This period usually extends for many years, beginning before and continuing after the last menstrual event.

The average age of menopause in North America is presently 51 years. Despite the fact that four out of five postmenopausal women have some symptoms of estrogen deficiency, it has been estimated that only 10 to 20% of them will seek medical attention.

Endocrinologic changes described above coincide with the development of certain clinical phenomena.

Clinical Findings Associated with Menopause

1. Vasomotor symptoms
2. Osteoporosis
3. Atrophy of the genitalia (atrophic vaginitis, dyspareunia)
4. Cardiovascular disease
5. Insomnia
6. Depression
7. Anxiety

A. Vasomotor Symptoms

1. Etiology

Perhaps the most dramatic symptoms in the menopausal woman include hot flushes and night sweats. Their etiology still remains a mystery. More than half of all postmenopausal women will develop these symptoms; their frequency and severity, however, are quite variable. Recent research suggests a link between central (CNS) neuroendocrine mechanisms responsible for the initiation of episodic LHRH release and the onset of flush episodes. They do not, however, implicate hormones directly as causing these symptoms.

2. Treatment

The most effective treatment for flushes is the administration of estrogens. Doses of up to 0.625 mg of conjugated estrogens (or equivalent) given cyclically (21 days per month) or with progesterone are effective in the vast majority of patients. Estrogens should probably be given at this dose for a few months; a gradual decrease in the dosage with eventual withdrawal can be achieved over a 1-year period with minimal symptom recurrence. Although vasomotor symptoms may not recur, long-term use of estrogens prevents many other postmenopausal problems.

In cases in which estrogens are contraindicated, progestogens may be used; results are usually less dramatic. Clonidine, in doses of 25 to 100 μg twice daily, can also provide symptomatic relief in more than 50% of cases.

Treatment of Menopause-Related Symptoms

Symptoms	Treatment
1. Hot flushes	a. Conjugated estrogens 0.625 mg or equivalent p.o. daily for approximately 6 months; gradual withdrawal over a 1-year period Or b. Progestogens: several preparations available; daily p.o. administration; for dosage consult individual drug literature Or c. Clonidine, 25 to 100 μg p.o. twice daily.
2. Atrophy of genitalia	a. Conjugated estrogens, 0.625 mg or equivalent p.o. 21 days per month for a few months Or b. Intravaginal estrogen preparations, 21 days per month for a few months
3. Osteoporosis	a. Estrogens (cyclic or with progesterone) b. Calcium c. Exercise as tolerated
4. Psychiatric symptoms (anxiety, depressions)	Estrogens not proven effective (no advantage over placebo); treat as in other patients

B. Osteoporosis

A detailed discussion of this subject can be found in Section I of this chapter. Briefly, estrogen depletion accelerates bone loss. Replacement therapy, when not contraindicated, should be routinely recommended in women considered at high risk of developing clinical osteoporosis.

C. Atrophy of the Genitalia

Decreased estrogen levels result in vaginal atrophy in the vast majority of postmenopausal women. Dyspareunia may result, especially in women who do not maintain regular sexual activity following menopause. In addition, trigonitis may result in symptoms of frequency, urgency, and hesitancy.

Estrogen administration is highly effective in reversing atrophic vaginitis, trigonitis, and associated dyspareunia. Intravaginal estrogen use, although effective, leads to circulating levels as high as these associated with oral estrogen administration.

D. Cardiovascular Disease

Although epidemiologic evidence suggests a sharp increase in the incidence of myocardial infarction and cerebrovascular disease in women after age 40, a cause-and-effect relationship between decreased estrogen levels and increased incidence of vascular disease has not yet been established.

Several studies have reported no effect of estrogen administration on the prevalence of coronary artery disease in postmenopausal women. Only one report claimed some benefit, and another suggested the opposite.

E. Psychiatric Symptoms

Although insomnia, depression, and anxiety may be chronologically related to menopause, they may or may not be caused by estrogen withdrawal. Although controversial, evidence suggest that estrogen therapy has no effect on psychological well-being.

F. Benefits and Risks of Estrogen Therapy

Benefits and risks of estrogen therapy are listed.

Benefits and Risks of Estrogen Therapy

Benefits

Proven

1. Control of vasomotor symptoms
2. Prevention of osteoporosis
3. Treatment of atrophic vaginitis/trigonitis

Questionable

1. Prevention of atherosclerosis
2. Treatment of depression, anxiety, insomnia

Risks

Well Documented

1. Endometrial carcinoma
2. Mild water retention with occasional mild hypertension

Undocumented

1. Hypercoagulability
2. Obesity
3. Cardiovascular diseases
4. Breast cancer

Since 1975, a wealth of data has accumulated regarding the risk of endometrial cancer in long-term users of estrogen therapy. Although evaluation of all the evidence is an impossible task, a summary of findings is presented.

1. There appears to be an increased incidence of endometrial cancer among estrogen users.
2. Relative risk appears to be higher in the United States than in most Western European countries.
3. The death rate from endometrial carcinoma is between 100 and 200 per million women over 55.
4. The relative risk of developing endometrial carcinoma appears to be proportional to the total duration and dosage of estrogen used.
5. Obese women have a higher risk of developing endometrial carcinoma.
6. Cyclical low-dose estrogen therapy is associated with a much smaller incidence of endometrial cancer.
7. Cyclical low-dose estrogen plus progestogen (for the last 7 to 10 days of each cycle) is not associated with a significant increase in the risk for endometrial carcinoma.
8. Cytological examination of small samples of the endometrium obtained by suction biopsy may not be adequate for the diagnosis of early-stage endometrial cancers.

It should be noted that unlike the oral contraceptive literature, no significant increase in thromboembolic episodes has been reported in postmenopausal women treated with estrogens. In addition, there are no proven serious risks associated with low-dose estrogen therapy following hysterectomy.

SUGGESTED READING

Davidson M. D.: The effect of aging on carbohydrate metabolism: A review of the English literature and a practical approach to the diagnosis of diabetes mellitus in the elderly. *Metabolism* 2:688, 1979.

Gambert S. R.: Hyperthyroidism in the aged. *Geriatr Consult* 3:16, 1984.

Gambert S. R.: Atypical presentation of thyroid disease in the elderly. *Geriatrics* 40:63, 1985.

Gambert S. R., Tsitouras P. D.: Effect of age on thyroid physiology and function. *J Am Geriatr* Soc 33:360, 1985.

Kannel W. B., Gordon T., Castelli W. P.: Obesity, lipids, and glucose intolerance. The Framingham study. *Am J Clin Nutr* 32:1238, 1979.

Lipid Research Clinics Program: The lipid research clinics coronary primary prevention trial results, I. Reduction in incidence of coronary heart disease, *JAMA* 23:351, 1984.

Wallach S.: Treatment of Paget's disease. *Adv Intern Med* 27:1, 1982.

8

Dermatologic Problems in the Elderly

I. INTRODUCTION

The skin is a heterogeneous organ that undergoes numerous changes with advancing age. Typically, the diminished reserve and functional capacity of the skin does not pose any life-threatening condition. Normal age-related changes, however, may lead to a wide variety of symptomatology as well as a change in cosmetic appearance.

Age-Related Changes in Skin: Clinical Implications

Associated Clinical Findings	Age-Related Change
Diffuse Atrophy	
1. Lax skin and wrinkling	1. Biochemical changes in elastin and collagen molecules lead to loss of normal function
2. Increased sensitivity to blistering forces	2. Effacement of dermal papillae and epidermal pegs with dermal thinning and a flat dermal–epidermal junction

---

(continued)

Age-Related Changes in Skin: Clinical Implications
(Continued)

Associated Clinical Findings	Age-Related Change
Diffuse Atrophy	
3. Increased numbers of "senile" purpura	3. Decreased numbers of dermal vessels, capillary loops, mast and Langerhans cells
4. Decreased resistance to infection	
5. Decreased sensitivity to allergens (contact dermatitis)	
Hyperpigmentation	
1. Macules	1. Macules (freckles or lentigines) in solar areas are one of the earliest signs of aging skin
2. Generalized hyperpigmentation in body areas (neck)	2. Decreased total melanocyte number but greater percentage of dopa-positive melanocytes in solar-exposed skin
Slower Wound Healing	
1. Slower nail growth	1. Decreased rate of turnover for keratinocytes
2. Slower hair growth	2. Smaller hair follicles with finer hair because of decreased androgen stimulation
Asteatosis and Comedones	
	1. Xerosis (dryness): legs > arms > face
	2. Decreased sebaceous gland function (no atrophy) with complete inactivity after age 70 because of changes in androgens

(continued)

Age-Related Changes in Skin: Clinical Implications
(Continued)

Associated Clinical Findings	Age-Related Change
Decreased Sweating	1. Atrophy of eccrine sweat glands and loss of innervation
Hair Graying	1. Onset genetically determined
Male Pattern Balding	1. Onset genetically determined
	2. Hairs change from terminal to velum

II. PRURITIS

One of the most common dermatologic complaints in the elderly is pruritis, typically caused by asteatosis. This problem presents most commonly along the anterior aspect of the tibia; erythema, shiny atrophic skin, and even excoriation may result. Treatment consists of less frequent bathing (water is drying to the skin), moisturizing creams, and appropriate care of any superimposed infection. Although antihistamines, neuroleptics, and other antipruritics are sometimes effective, side effects must always be carefully considered.

Common Causes of Pruritis in the Elderly

Asteatosis (xerosis)
Contact dermatitis
Atopic dermatitis
Psoriasis
Lichen simplex chronicus (neurodermatitis)
Drug reactions

(continued)

Common Causes of Pruritis in the Elderly (*Continued*)

Hormonal (thyrotoxicosis)
Metabolic (uremia, hepatic failure with cholestasis)
Psychiatric delusions (parasitosis)
Malignancy (mycosis fungoides, lymphoma, carcinoma)

III. DISEASES OF THE SKIN

Aging, genetic predisposition, and cumulative sun exposure
lead to an increased prevalence of many benign and malignant skin
conditions in the elderly. The exact contribution of each factor is
difficult to evaluate even in studies using identical twins. Although
genetic information apparently determines the likely density and
location of lesions such as seborrheic keratosis, lentigines, and
actinic keratosis, sun exposure can modify these. The prevalence of
dermatologic lesions in the elderly is very high, with estimates
ranging from 60% of the general elderly population to 95% of
hospitalized elderly.

Skin Diseases with Increased Prevalence in the Elderly

Benign	Vascular ectasias
	Pedunculated fibromas
	Actinic keratosis
	Leukoplakia
	Keratoacanthoma
	Stasis dermatitis and ulcers
	Bullous pemphigoid
Infectious	Candidiasis
	Herpes zoster
	Decubitis ulcers (frequent superinfection)
Malignant	Bowen's disease
	Basal cell epithelioma
	Squamous cell carcinoma
	Lentigo maligna/malignant melanoma
	Metastatic carcinoma: breast, lung, GU, stomach, colon, kidney, lymphoma, leukemia

---

Benign proliferations are most common. **Vascular ectasias** consist of cherry-colored papular angiomas that appear in both sun-exposed and nonexposed areas; **telangiectasias** are related to prior actinic exposure; and **blue-hued venous lakes** occur on ears or other facial areas. **Skin tags** and **fibromas,** the former skin colored and the latter dark reddish-brown papules, are managed by simple scalpel removal. Premalignant lesions include actinic **keratosis, leukoplakia,** and **keratoacanthomas.** The latter has been considered by some a squamous cell epithelioma in regression and consists of a firm raised nodule with a central ulcerated crater. Surgical removal is recommended. **Leukoplakia** is a flat white plaque on oral, vaginal, or vulval mucosa and has a 20–30% progression rate to squamous cell carcinoma. Etiologic factors that have been implicated include chronic inflammation (cheilitis from ill-fitting dentures), chronic infection (vaginitis), alcohol, smoking, and sunlight. Treatment consists of close inspection with biopsy of any area suspicious for malignancy, avoidance of causative factors, and electrosurgery for any persistent lesions.

Actinic keratoses are flat papules firmly adherent to the skin. They occur most commonly in sun-exposed areas, typically in fair-complexioned, blue-eyed persons. Color may appear tan, brown, or reddish. A mild itch may cause patients to pick at the lesion. Rarely, a cutaneous "horn" develops, presenting a very pro-liferative form of the lesion; this form has a higher potential for malignant transformation and deserves immediate attention. Management consists of biopsying any rapidly growing, thickened areas to rule out any progression to squamous cell carcinoma. Fortunately, this transformation occurs in only a small percentage of cases. Sun protection with UV sunscreen cream and use of hats, long-sleeved garments, and scarves should be encouraged. Individual lesions may be treated with curettage or topical liquid nitrogen application. Numerous lesions can be treated with 5-fluorouracil cream twice a day over a 2-week period of time (if the patient can tolerate an entire course) with repeat treatment as necessary. Follow-up is mandatory.

Numerous other benign lesions occur with increased prevalence in the elderly.

Stasis dermatitis and **cutaneous ulcers** may develop in persons with impaired venous return from the lower extremities, resulting in constrictive tissue swelling and local arterial insufficiency. The typical patient suffers from venous varicosities, congestive heart failure, obesity, or prolonged upright posture. Primary sites include the medial or lateral malleoli and along the anterior tibia.

The lesion is usually brawny colored, resulting from hemosiderin deposition in cutaneous tissues. Induration, lichenification, and frequently central ulceration may be present. These are shallow, punched-out lesions with sharp margination. Symptomatology may include pruritis, aching, or burning sensations. Treatment includes soaking and debridement, elevation of legs to the level of the heart, antibiotics for any superinfection, and measures to prevent reaccumulation of stasis swelling: support stockings or zinc oxide (Unna) boots. Although topical steroids may be of use in acutely inflamed cases, they should be avoided chronically. Skin grafts may occasionally be curative.

Bullous pemphigoid is largely a disease of persons over 60 characterized by large, tense, subepidermal bullae. Anti-basement-membrane antibodies can be detectable by immunofluorescence studies on skin biopsy samples. Rarely, bullae may occur on mucous membranes. The disease usually follows a remitting, relapsing course. Mild cases may be treated with topical steroids and warm compresses; more severe cases respond well to systemic corticosteroids, immunosuppressives, or antimalarials.

Infectious dermatologic conditions are more common in geriatric patients because of the thinned barrier and decreased vascularity of the skin, providing fewer local immune effector cells. Pathogens capable of causing erysipelas and impetigo can be life threatening. **Candida** species can be isolated more frequently from the skin of elderly patients and are likely to cause symptomatology in the form of thrush, perleche, vaginitis, and paronychia. Predisposing conditions common to the elderly include diabetes mellitus, systemic antibiotic, steroid, or immunosuppressive therapy, macerated tissue, and malnutrition. Treatment consists of drying the skin, removing any predisposing condition, and appropriate antibiotic or antifungal therapy (nystatin, miconazole, or clotrimazole) when indicated.

Herpes zoster (shingles) represents a recrudescence of herpes varicella/zoster virus latent in the dorsal root ganglion. Reactivation can be triggered by trauma, stress, T-cell immunosuppression, or illness. A prodrome of tingling, itching, or burning is common. The eruption covers a unilateral dermatome and consists of clear vesicles. Two-thirds of cases occur in those over 50 years of age; affected younger persons are usually immunosuppressed. Treatment is symptomatic with drying lotions and pain medication unless the trigeminal nerve is involved. In this case, keratitis could develop requiring opthalmologic evaluation and corticosteroid drops.

Zovirax may replace adenosine arabinoside as the treatment of choice for dissemination outside the dermatome or in cases resulting in pneumonia, hepatitis, or CNS involvement. Complications include postherpetic neuralgia in over half of those affected patients over 65 years of age. This is a dermatomal pain persisting longer than 4 weeks from the resolution of skin lesions. Preventative steroid therapy is indicated in this age group if not immunosuppressed. Doses of 60 mg/day oral prednisone for 5–7 days, followed by a rapid taper over 2–3 weeks, are usually effective in reducing symptoms. Severe resistant postherpetic neuralgia that does not respond to prednisone may require neurological and/or neurosurgical evaluation.

IV. DECUBITUS ULCERS

Decubitus ulcers are another form of ischemic ulcer leading to tissue necrosis. The presence of a decubitus usually signifies a medical illness leading to immobility, inability to sense ischemic pain, poor tissue integrity, or skin maceration. Superinfection is common; osteomyelitis in adjacent bone, recurrent fevers, or frank sepsis may also result. Treatment includes relief of local pressure and wetness, antibiotic treatment of infections, thorough cleansing, debridement (either mechanical or chemical) and decontamination, and promotion of granulation by moistened protective gauze dressings. Skin grafting may be appropriate in noninfected advanced cases.

Decubitus Ulcers: Etiology and Associated Problems

Etiology	Associated Problems
1. Hypoxemia	Local unrelieved pressure (immobility or neuropathy)
	Anemia
	Local atherosclerotic disease
	Hypoxia

(continued)

Decubitus Ulcers: Etiology and Associated Problems (*Continued*)

Etiology	Associated Problems
2. Maceration	Urinary or fecal incontinence, moist bedclothes Insufficient debridement of necrotic tissue
3. Poor tissue integrity	Malnutrition Deficiency of trace metals (Zn) or vitamins Malignancy

Decubitus Ulcers: Clinical Characteristics

Clinical Staging
I. Mild erythema or skin atrophy
II. Simple skin break
III. Dermis involved in ulcer
IV. Ulcer penetrating through fascial planes

Bacteriology of Superinfection
Common: *Staphylococcus, Streptococcus,* aerobic gram-negative bacilli
Rarely: Anaerobes, enterococci, *Pseudomonas*

Complications
Recurrent fevers
Anemia of chronic disease
Sepsis
Adjacent osteomyelitis

Treatment
a. Mechanical or chemical debridement as necessary; a skin graft may be necessary
b. Frequent movement, water or air mattress, repositioning q2h to prevent pressure
c. Decontamination: H_2O_2 : iodine solution (redness may obscure erythema from infection)
d. Moist, O_2-permeable, protective dressing until granulation occurs (saline, Ringer's lactate); dry dressings may remove existing granulation tissue

V. MALIGNANT SKIN LESIONS

Malignant lesions of the skin have a prevalence of approximately 15/100,000 in Americans over 65 years of age. The most common lesion in this age group is **squamous cell carcinoma** with a prevalence of 10/100,000. This presents typically in sun-exposed areas in fair-haired, blue-eyed individuals. The lesion may be a verrucous plaque or nodule with a raised border. The color ranges from skin toned to reddish brown, with crusts and ulcerations common. **Bowen's disease** is an intraepidermal squamous cell carcinoma *in situ,* which appears as a reddish-brown scaly plaque in non-sun-exposed areas. Biopsy for diagnosis and complete surgical excision are mandatory for both. Cancers arising in scars, sites of chronic inflammation, or radiodermatitis tend to be more invasive.

Basal cell epitheliomas are slow growing pearly nodules in sun-exposed skin. A rolled border can usually be detected. Telangiectasias may appear over the surface, or ulceration may be present. Treatment following biopsy diagnosis involves surgical removal followed by electrocautery and desiccation. Rarer forms include an erythematous plaque with a rolled border and a scarlike shiny telangiectatic skin area (morpheaform basal cell epithelioma); both are locally invasive.

Malignant melanoma is a malignant transformation of melanocytes with potential for metastatic dissemination. A prevalence of 1/100,000 in persons over age 65 has been reported. Four histological types are recognized; the lentigo maligna type is most common in the elderly. Initially, the tumor undergoes a stage of horizontal growth during which the tendency to metastasize is low. Since this lesion tends to progress to vertical growth with invasion, it should be removed with adequate margins to the level of subcutaneous fat. Warning signs in a preexisting nevus include changes in size, color, texture, margins, pruritis, bleeding, or ulceration. Lesions with a deep level of invasion (Breslow's lesional depth), proximal location, and vascular invasion have a poorer prognosis. A lesion <0.75 mm in size is associated with a 95% 5-year survival rate. On the other hand, the 5-year survival with subcutaneous invasion is approximately 20–30%. Since malignant melanoma accounts for 60–70% of all deaths attributable to skin cancer, early detection and appropriate removal are essential. Moh's chemosurgery, i.e., evaluating the base of the excision site by frozen section to include any deeply invasive "roots" in the resection, has given the hope of improving cure rates. Metastatic disease is best treated using combined chemotherapy protocols and/or immunotherapy.

Melanoma in the Elderly: Classification and Staging

Type	Description	Potential
1. Lentigo maligna	Atypical melano-cytes spread over a large skin area; var-iegated color with advanc-ing/receding margin	One-third to one-half pro-gress to in-vasive melanoma
2. Nodular	Rapidly growing blue-black nod-ule; absent hair follicles on lesion	100% invasive
3. Superficial spreading	Elevated, ar-ciform lesion with notched edges; varie-gated colors: black, red, blue, brown, white; satellite lesions fre-quent	100% invasive
4. Acral lentiginous	Flat, variegated lesions on dis-tal extremities: subungal, in-terdigital, mucous mem-branes	100% invasive; better prog-nosis

Clinical Staging: Clark's Levels

I	Intraepidermal melanoma
II	Invasion of papillary dermis

(continued)

Melanoma in the Elderly: Classification and Staging
(Continued)

Clinical Staging: Clark's Levels (Continued)
III   Filling but not exceeding papillary dermis
IV   Invasion of reticular dermis
V   Invasion of subcutaneous tissue

Clinical Staging: Breslow's Lesional Depth
Good prognosis   Depth of 0.75 mm or less
Poor prognosis   Depth 0.76 mm and deeper,
                  progressively poorer
                  prognosis

Clinical Staging: Location of Lesion
Good prognosis   "BANS" lesions: upper back,
                  upper posterolateral arm, pos-
                  terolateral neck, posterior
                  scalp
Poor prognosis   Non-"BANS" lesions

VI. CANCER AND SKIN

Metastases to the skin are not an uncommon finding. It has
been estimated that as many as 5% of malignancies metastasize to
the skin as the primary site of spread. Most common affected sites
include skin of the head, neck, chest, and abdomen. In women,
breast, colon, and lung carcinomas are the most common
responsible primary lesions; in men, lung, colon, and head and
neck tumors are most common. Although tumors usually metastasize near the primary lesion, renal cell carcinomas tend to metastasize to the skin of the head, and breast carcinomas most commonly
to the scalp. Metastatic lesions to the skin are typically firm,
subcutaneous, erythematous nodules; plaquelike lesions are often
found in cases of breast carcinoma. Multiple sites of involvement
are more commonly seen in cases of gastric carcinoma.

In addition to direct invasion of the skin, several dermatologic
syndromes have been associated with internal malignancies.

Skin Changes with Internal Malignancy

Dermatologic Finding	Malignancies to Be Considered
Exfoliative erythroderm	50% association with lymphoma if onset noted after age 50; leukemia, lung, and rectal carcinoma
Sudden-onset vitiligo	Melanoma, gastric carcinoma
Dermatomyositis: heliotrope erythema of eyelids, red-blue plaques of knuckles, alopecia, or scalp erythema	Lung, GI, breast, lymphoma, renal cell carcinoma
Acanthosis nigricans: hyperpigmentation with hyperkeratosis in skin folds	Gastric, pancreas, rectal, colon, lung, ovary, hepatic, breast, uterine carcinoma
Thrombophlebitis migrans: successive tender nodules over veins in all extremities	Adenocarcinoma (lung, pancreas, colon, gastric)
Sudden-onset seborrheic keratoses: multiple lesions, Leser–Trelat sign	Gastric carcinoma
Paget's disease: gray to erythematous plaque with scaling, crusts, or erosions	Breast, rectal, vulvar carcinoma

SUGGESTED READING

Aubry T., MacGibbon B.: Risk factors of squamous cell carcinoma of the skin: A case control study in the Montreal region. *Cancer* 55:907–911, 1985.

Carter D. M., Balin A. K.: Dermatologic aspects of aging. *Med Clin North Am* 67(2):531–543, 1985.

Johnson O. K., Emrich L. J., Karakouisis C. P., et al: Comparison of prognostic factors for survival and recurrence of malignant melanoma of the skin, clinical stage I. *Cancer* 55:1107–1117, 1985.

Proper S., Fenske N.: Common skin tumors in the geriatric population (pictorial). *Geriatr Med Today* 4(9):17–33, 1985.

Seiler W. O., Stahelin H. B.: Decubitus ulcers: Treatment through five therapeutic principles. *Geriatrics* 40(9):30–42, 1985.

9

Hematologic Diseases in the Elderly

I. INTRODUCTION

Although hematopoietic and immunologic parameters undergo a myriad of changes with age, the majority of these changes are clinically insignificant. As this cell population is exquisitely sensitive to changes in general health, surveys attempting to establish "normal" values by screening large numbers of geriatric patients must stringently exclude intercurrent illness.

II. RED CELL DISORDERS

Anemia is the most common hematologic disorder in the elderly. Although often asymptomatic, it can be particularly problematic in elderly with emphysema, coronary artery disease, and other chronic disorders. There is still debate as to what levels of hemoglobin and hematocrit warrant evaluation. By standard criteria, 21% of elderly women and 34% of elderly men are reportedly anemic. In 1978, the Department of Health, Education, and Welfare conducted the National Health and Nutrition Evaluation Survey II (NHANES II). In this study, 15,000 persons were selected after physical examination and history to represent the "healthy" American population. Pregnant women and patients with hemoglobinopathies were excluded. Ninety-five-percent confidence limits and median hemoglobin values were determined in men and women.

Blacks had slightly lower values at all ages than whites. The values
for men declined with increasing age; although unclear, it was
speculated that this resulted from changes in androgenic stimulation
of bone marrow. After age 60, there was a slight sex difference in
hemoglobin values, with mean values for men still slightly higher
than for women.

Hemoglobin in Healthy Adults over 65 Years of Age

Hemoglobin (g/dl)			
Women	Men	Sources	Comments
≥12	≥14	Standard criteria	30% of elderly men labeled "anemic"
≥12	≥13	WHO criteria	World Health Organization standards
11.7–16.1	12.6–17.4	NHANES II	1978: 95% confidence limits

Age-Related Hematologic Changes

Parameter	Comment
↑ MCV to a median of 89	No change in MCHC as cellular Hb content rises proportionally
↓ RBC survival	Clinically insignificant
↓ Marrow cellularity	1–30 years: 50% decrease from birth noted
	31–69 years: stable cellularity
	Over 70 years: 40% decline from adult levels
	Etiology:
	No reproducible drop in erythropoietin
	Not a stem cell defect or deficit (infused stem cells do not correct cellularity)
	Probably ↓ colony-forming unit–erythrocyte factors from monocytes (CFU-E)

(continued)

Age-Related Hematologic Changes (*Continued*)

Parameter	Comment
Myeloid/erythroid ratio	Preserved
↓ Colony stimulation of stem cells	Stress conditions
Defective iron utilization	↑ Marrow iron
	↑ Serum ferritin (reflecting stored iron)
	↓ Serum iron (from normal 115–130 to 70–80 µg/dl)
	Percentage transferrin saturation unchanged at 25–30% as TIBC decreases

Although causes of anemia do not change with age, in younger persons different relative frequencies are noted. Iron deficiency in the young usually reflects heme loss from pregnancy and menstruation; in the elderly, it typically reflects a loss from the GI tract in both sexes. Although congenital hemolytic anemia and hemoglobinopathy may still be present in the 20- to 29-year-old population, only the rarer acquired hemolytic anemias are present in the elderly.

Causes of Anemia by Age

	Age (years)		
Etiology	20–29	40–49	>60
Iron deficiency	20.6%	10.1%	12.3%
Megaloblastic anemia	1.0%	0.7%	3.2%
Hemolytic anemia	1.3%	1.0%	1.0%
Aplastic anemia	3.3%	4.0%	0.7%
Hematologic malignancy, including preleukemia	13.9%	5.3%	2.7%
Anemia of chronic disease	63.3%	71.9%	81.2%

If anemia of chronic disease is further broken down to exclude chronic liver and renal disease, it is felt to be responsible for 48.0, 56.0, and 75.0% of cases of anemia in those 20–29, 40–49, and greater than 60 years of age, respectively.

Anemia in the elderly can be classified by marrow functioning. Hypoproliferative anemias may be normocytic/normochromic or hypochromic/microcytic; anemia from ineffective erythropoiesis is often megaloblastic; and hyperproliferative anemias are either normocytic/normochromic (blood loss) or show atypical forms. (See table on p. 211.)

III. ANEMIA OF CHRONIC DISEASE

Anemia of chronic disease is the most commonly diagnosed form of anemia in the elderly. Even excluding chronic renal and liver disease, anemia of chronic disease accounts for 75% of all anemia in elderly adults. A number of chronic conditions can produce an anemia of chronic disease. The anemia in all of these conditions, however, improves with treatment of the underlying condition.

Anemia of Chronic Disease: Etiology

Autoimmune, collagen vascular disease:
  Rheumatoid arthritis, lupus, polymyositis, inflammatory
    bowel disease, thyroiditis
Chronic renal insufficiency
Chronic hepatic insufficiency:
  Hepatocellular, cholestatic
Endocrinopathy:
  Hypopituitarism, hypothyroidism, Addison's
Nonhematologic malignancies
Chronic infections:
  TB, Fungal
  Osteomyelitis
  Pyelonephritis
  Subacute bacterial endocarditis
  Diverticulitis/perirectal abscess
  Tissue/decubiti infections
  Chronic sinusitis
  Dental abscess
Chronic inflammation:
  Sarcoid
  Psoriasis

Classification of Anemias in the Elderly

Hypoproliferative (low reticulocyte count)	Ineffective erythrocytosis (low reticulocyte count)	Hyperproliferative (high reticulocyte count)
1. Anemia of chronic disease	1. B₁₂ deficiency	1. Blood loss
2. Iron deficiency anemia	2. Folic acid deficiency (or antimetabolite chemotherapy)	2. Hemolytic anemia
3. Aplastic anemia	3. Sideroblastic anemia (rarely pyridoxine responsive)	3. Hereditary spherocytosis
4. Hematologic anemia	4. Refractory anemia/myelofibrosis	4. Thalassemia minor

Characteristics of Anemia of Chronic Disease: Diagnosis

Hypochromia on smear examination (earliest finding)
Mild, nonprogressive anemia (Hb 10–12)
Typically normocytic; may be microcytic (smokers)
Inappropriately low corrected reticulocyte count
No additional cytopenias
Evidence of iron-deficient erythropoiesis: TIBC >250 μg/dl
    with MCV <80 or transferrin saturation <20%
Evidence of adequate iron stores: Marrow iron present or
    ferritin >50 μg/liter
Response to treatment of underlying condition

Ferritin may be artificially elevated in an iron-deficient patient as a result of inflammatory conditions; the expected TIBC rise may be suppressed in protein malnutrition, liver disease, or other chronic conditions suppressing protein synthesis. Two methods have been suggested as a way to help distinguish between an iron deficiency anemia and anemia of chronic disease. A trial of oral ferrous iron may be attempted (1 g/day). A greater than 3% rise in Hct after 6 weeks of therapy supports the diagnosis of a preexisting iron deficiency. Alternatively, a marrow examination may be performed to demonstrate the presence or absence of stainable iron. Although often harder to obtain, this latter approach has the advantage of being definitive. If iron is present, an investigation for the underlying chronic disease is warranted, and unnecessary long-term iron therapy can be avoided. It also provides a screen for less frequent marrow conditions causing a low-reticulocyte, normocytic anemia and defective erythropoiesis. Any prolonged drop in Hb below 10 mg/dl in the absence of intercurrent infection or stress should prompt a marrow examination. Other signs suggesting the need for marrow examination include the presence of additional cytopenias or abnormal cellular forms on examination of the peripheral smear.

IV. OTHER CAUSES OF CHRONIC ANEMIAS WITH HYPOPROLIFERATIVE OR INEFFECTIVE ERYTHROPOIESIS

Anemia	Comments
Aplastic anemia	Hb <10
	Other marrow elements depressed as well
	Implicated drugs: gold, phenytoin, phenobarital, chloramphenicol, phenylbutazone
	May respond to drug withdrawal, androgens, antithymocyte, globulin
	Pure red cell aplasia may accompany a thymoma
Sideroblastic anemia	Mitochondria containing nonutilized iron accumulate around nucleus, creating the "beaded circle"
	Average indices often are normal; smear may show markedly hypochromic cell population
	Implicated drugs: INH, chloramphenicol, lead, EtOH
	Rare cases respond to drug withdrawal, or a pyridoxine trial
	Iron therapy contraindicated
	10% progress to acute myelogenous leukemia (AML)
Myelofibrosis	Hb <10
	50% have thrombocytopenia
	Myelophthisic smear (poikilocytes, nucleated RBCs)

(continued)

Anemia	Comments
Hairy-cell leukemia	Hypercellular marrow 20–50% progress to AML within 1–10 years
	Anemia presents first; later, neutropenia/thrombocytopenia seen
	Lymphocytes with spiculated projections, movement on phase microscopy, (+)tartrate-resistant acid phosphatase stain
	70% have splenomegaly
Refractory anemia (RA)	Hb <10
Refractory anemia with excess blasts (RAEB)	50% neutropenia
	<5% RA and >8% RAEB exhibit marrow blasts
	15% of RA and 30% RAEB progress to AML

V. IRON DEFICIENCY ANEMIA

Iron deficiency anemia is the second most frequent cause of anemia in the elderly. Despite the traditionally limited diets of many adults, dietary deficiency is extremely rare; iron reutilization is extremely efficient. Normal body stores of heme iron are sufficient (100 mg), and dietary intake must only replete the equivalent iron lost normally in the GI tract each day (2 cc blood loss). This latter amount is equivalent to 1 mg heme iron absorbed or 10 mg of dietary iron ingested. Meat, eggs, and cereal are good sources of dietary iron. Deficient diets, however, may lead to iron deficiency after many years. The presence of ascorbic acid in the diet, adequate dietary iron, normal gastric acidity, and an iron-deficient marrow all enhance iron absorption. Even with adequate iron intake, however, elderly persons with achlorhydria will not be able adequately to convert ferric iron (dietary form) to ferrous iron (absorbable). This also will predispose to a deficit anemia.

In all cases of iron deficiency anemia, blood loss must be ruled out. The most common site is gastrointestinal, and the patient may only be intermittently guiac positive from a chronic, low-grade bleed. Therefore, inability to document guaic positivity should not postpone a GI evaluation for the site of blood loss in an iron-deficient geriatric patient.

Iron Deficiency in Elderly Adults: Etiology

Gastrointestinal	Colonic telangiectasias
	Diverticuli
	Adenocarcinoma
	Fissure
	Gastritis
	Hemmorrhoids
Extravascular	Surgery/trauma
	Phlebotomy
Gynecological	Dysfunctional uterine bleeding
	Leiomyomata
	Carcinoma
Renal	Chronic cystitis
	Carcinoma

Decreased heme synthesis results from iron deficiency and leads to the formation of microcytic cells. Serum ferritin falls, reflecting decreased iron stores. It is important to remember, however, that ferritin may be nonspecificially increased in hepatic disease, chronic inflammatory conditions, and other causes of disordered erythropoiesis. Since erythrocytes are synthesized without complexed iron in heme, the content of free protoporphyrin rises. In early cases of iron deficiency with one or more of the above indices abnormal (ferritin, transferrin saturation, or free erythrocyte protoporphyrin), a marrow examination will reveal no stainable iron in at least 50% of cases even before anemia is observed. Treatment of iron deficiency requires oral iron in the ferrous state presented in high excess (1 g/day). In most cases any absorptive problems can be overcome; parenteral iron products lead to reticulocytosis only 24–48 hr earlier, involve increased risk of hypersensitivity, and are painful.

Iron Deficiency Anemia: Diagnosis

1. Microcytosis (MCV <80) develops first
2. Hypochromia/anemia
3. Transferrin saturation <16% (16–20% indeterminate) and/or
   Ferritin <12 μg/liter (12–50 μg/liter indeterminate) and/or
   Free erythrocyte protoporphyrin >75 g/dl
4. Absent marrow iron

Other rare causes of microcytic anemia that may escape detection until old age include thalassemia minor and hereditary spherocytosis. Although the degree of anemia may be mild and not have caused symptomatology previously, an eldery person may poorly tolerate these conditions. Both have elevated reticulocyte counts and abnormal smears revealing either nucleated red cells and target cells or typical spherocytes

VI. MEGALOBLASTIC ANEMIA

Megaloblastic anemia develops when cytoplasmic growth continues despite arrest of nuclear maturation. This results from restricted nucleic acid synthesis. Any defect in nuclear maturation can lead to megaloblastosis, as occurs in antimitotic chemotherapy. Most commonly, this condition results from a folate deficiency. Leafy green vegetables, fruits, and nuts are good dietary sources of folate; dietary lack usually leads to folate deficiency within 2–3 months. Certain medications may also impair folate metabolism and must be looked for. Since this form of anemia often fails to occur despite a diet deficient in folate, nondietary factors are clearly important in manifesting this form of anemia.

Megaloblastic Anemia

Megaloblastic erythrocytes (MCHC >33)
Multilobulated neutrophils (any six-lobed; or three-five-
   lobed)

Etiology
   Antimitotic chemotherapy
   Liver disease
   Nutritional deficiency: Folate
   Alcohol usage
   B_{12} deficiency

Drugs decreasing folate absorption
   INH
   Phenytoin
   Phenobarbitol
   Cholestyramine

Drugs antagonizing folate action
   Alcohol
   Methotrexate
   Trimethoprim, other antibiotics

Because serum folate (abnormal <3 mg/ml) and RBC folate
levels can be affected by drugs and fluctuate quickly after dietary
intake, they do not accurately reflect chronic deficiency. A trial of
only 50 µg folate daily will lead to reticulocytosis in cases of
deficiency. Treatment with folate in excess at 1 mg t.i.d. for 2–4
weeks will overcome most problems with absorption.

Vitamin B_{12} deficiency increases in prevalence with age
through an accumulation of causative conditions. The average age
at onset is 60, when an incidence of 9/100,000 is noted in persons
of northern European extraction. A dietary deficiency is extremely
rare, as total body stores equal 5000 mg and the total daily
requirement is only 1–3 mg. A person adhering to a strict vege-
tarian diet, excluding all red meat sources of B_{12}, would still
require up to 10 years to develop a B_{12} deficiency. Deficiency
develops much sooner (3–5 years) when intrinsic factor (IF) is
deficient. Five milligrams of B_{12} per day undergoes enterohepatic
circulation and requires IF to be reabsorbed.

B_{12} Deficiency: Etiology

	Shilling test			
	Part I	Part I*	Part II	Part III
Pernicious anemia (anti-intrinsic-factor antibodies)	+†	+	–	–
Gastrectomy	+	–	–	–
Achlorhydria/H_2 receptor antagonists	+	–	+	+
Atrophic gastritis (parietal cell autoantibodies)	+	+	+	+
Pancreatic insufficiency	–‡	–	–	–
Ileal resection/ inflammatory disease	+·	+	+	+
Ileal bacterial overgrowth	+	+	+	–
Vegetarian diet	–	–	–	–

*B_{12} presented free, not bound to egg whites.
†Abnormal.
‡Normal.

Serum B_{12} levels of less than 100 mg/dl are virtually diagnostic: urinary methylmalonic acid levels are a more precise method in unclear cases. Levels between 100 and 200 mg/dl are indeterminate, with very few progessing from these levels to true deficiency. Evaluation of B_{12} deficiency requires the Shilling test to elucidate the site of abnormality. Radiolabeled B_{12} is given orally (either bound to egg whites for palatability or free), and i.m. B_{12}, 5000 µg, is given 1–2 hr later to saturate plasma binding sites. A 24-hr urine collection is obtained, and the percentage of radioactive B_{12} excreted is reported; normal values are at least 10% in most laboratories. Part II repeats the initial collection giving both radiolabeled B_{12} and intrinsic factor orally; part III repeats the entire part I after a 2-week course of tetracycline, 250 mg q.i.d. by mouth. No improvement in any section of the test should be expected in cases resulting from a resected or diseased ileum. Signs and symptoms of B_{12} deficiency in geriatric patients are numerous. It is important to remember that neurological signs and/or dementia may precede the onset of anemia.

B_{12} Deficiency: Signs and Symptoms

Symptoms	Signs	Laboratory
Fatigue	Weight loss	Megaloblastosis
Burning/painful tongue	Yellowish pallor	Anemia
GI complaints/ anorexia	Glossitis	↑ Bilirubin
Poor "balance"	Peripheral neuropathy (position and vibration)	↑ LDH
Weakness/inability to walk		Thrombocytopenia
Distal numbness/ tingling	Spasticity	Mild lymphocytosis
Confusion, psychosis, depression	Ataxia Disorientation	Gastric atrophy (by endoscopy)

Treatment includes 1000 μg i.m. cobalamin daily for 5–7 days and then monthly maintenance injections of the same dose. Rarely, an elderly patient will present with severe anemia (Hb <8) and cardiovascular compromise, necessitating an acute replacement of B_{12} with transfusions of packed red cells and judicious use of diuretics. Correction of marrow megaloblastosis is noted within 8 hr of therapy. Inappropriate treatment with folate may correct marrow defect, however it will not alter neurological sequelae of B_{12} deficiency.

VII. HEMOLYTIC ANEMIA

Hemolytic anemia presents typically with a high reticulocyte count and, when compensated, a nonprogressive anemia. The reticulocyte count may be blunted in cases with concomitant nutritional deficiency. Nonimmune causes for hemolysis include cell destruction because of membrane fragility or lack of malleability (paroxysmal nocturnal hemoglobinuria; spherocytosis) or conditions fragmenting normal cells (hypersplenism). The conditions associated with immune hemolysis are numerous; as many as 50% will remain idiopathic. (See table on p. 220.)

Immune Hemolysis: Etiology

Neoplasia	Collagen vascular	Infections	Drugs	Miscellaneous
Lymphoma	Systemic lupus erythematosis	Mycoplasma	α-Methyldopa	Ulcerative colitis
Chronic lymphocytic leukemia	Periarteritis nodosa	Syphilis	DOPA	Thyroid disease
Multiple myeloma	Rheumatoid Arthritis	Malaria	Penicillin	
Thymoma; teratoma		Viral: CMV, EBV, influenza		
Chronic myelogenous leukemia		Hepatitis A/B Coxsackie		

The innocent bystander phenomenon occurs when a drug–
plasma protein complex stimulates antibody production. The resulting antigen–antibody complex binds complement and leads to
nonspecific local hemolysis. Penicillin is an example of a drug that
is specifically bound to red cells. In patients who have been
previously sensitized, the clearance of complexed cells by splenic
macrophages may lead to a slow extravascular loss of red cells.
(See table on p. 222.)

Idiopathic autoimmune hemolytic anemia (IAHA) may occur
with either an IgG antibody in "warm hemolysis" (anti-IgG direct
Coombs +) or an IgM antibody in "cold hemolysis" (anti-C3
direct Coombs +). Although warm hemolysis typically occurs in
women aged 30–60, it may result from a specific drug reaction to
α-methyldopa in the elderly. Remission can usually be accomplished with corticosteroids and splenectomy. Danazol, 200–400
mg/day, as maintenance therapy may be more effective than
prednisone at preventing relapse. Cold agglutinin disease is usually
a disease of men over 60 and is precipitated by an antibody binding
maximally at 4°C. The antibody must acquire some thermal stability
and a high titer (cold agglutinin titer ≥1 : 1000) to cause clinical
symptoms. Patients present during winter with Reynaud's phe-
nomenum or acral cyanosis. Complement components are left on
the red cell surface, and cells are cleared by intrahepatic Kupffer
cells and splenic sinusoidal macrophages. Splenectomy is usually
ineffective as are coricosteroids; avoidance of cold, chronic folic
acid supplementation, transfusion of warmed cells when necessary,
and occasionally immunosuppressive therapy are the most effective
measures.

VIII. RED CELL DISORDERS

Polycythemia is defined as an increase in circulating hemoglobin above the individual's appropriate level, reflecting a true
increase in circulating red cell mass. Typically in the elderly the
95% confidence intervals have maximum levels of 16 g/dl for
women and 17 g/dl for men. True polycythemia must first be
differentiated from spurious polycythemia, an increase in hemoglobin circulation as a result of a contracted plasma volume. Both
cigarette smoking and chronic stress may lead to decreased plasma
volume for unclear reasons. True polycythemia may be caused by

Immune Hemolysis: Characteristics

Hemolysis type	Innocent bystander	Hapten	Autoimmune hemolysis	Cold agglutinin
Antigen	Drug	Drug	RBC protein	i antigen
Implicated drugs	Sulfa, quinidine, anti-TB, phenacetin	PCN, synthetic pen-icillins (Keflex)	Aldomet	
			RBC Rh locus	
			PCN, α-methyldopa	
Direct Coombs				
anti-IgG	−	+	+	−
anti-C3	+	−	−	+ (weak)
Indirect Coombs				
With drug	+	+	+	+
Without drug	−	−	+	+
Degree of hemo-lysis	Mild to severe	Absent to mild	Absent to se-vere	Absent to mild

hypoxemia, erythropoietin overproduction, or autonomous red cell proliferation.

Hypoxemia caused by various changes in oxygen delivery is most common. The true arterial oxygen saturation falls with abnormal hemoglobin structure, elevated altitude, permanent heme complexing (carboxy- or methemoglobin), or a change in the oxygen dissociation curve (acidosis). These factors are presumed constant in traditional blood gas calculation of oxygen saturation and should be independently considered in a polycythemic patient. An intermittent drop in saturation during physical activity or sleep apnea will not be evident on a resting blood gas determination taken while awake. Symptoms of polycythemia develop in smokers and patients with COPD because of a combination of carboxyhemoglobin, contracted plasma volume, and intermittent desaturation. The symptoms usually respond to phlebotomy and cessation of smoking. (See table on p. 224.)

Independent erythropoiesis, polycythemia vera, is a chronic myeloproliferative disease involving all marrow cell lines. Incidence increases dramatically after age 50, with an associated average survival of 10–12 years. A marrow examination is not mandatory; it typically reveals cellular hyperplasia. Early cases who do not fit established diagnostic criteria should be conservatively managed with phlebotomy until the diagnosis of polycythemia vera is confirmed or a secondary cause for polycythemia becomes apparent.

Diagnostic Criteria for Polycythemia Vera

Diagnosis Requires All of

A. Red cell volume increased:
   males >36 ml/kg or females >32 ml/kg
B. Arterial oxygen saturation normal:
   92% or higher
C. Splenomegaly

If Splenomegaly Is Absent, Any Two of the Following May Be Substituted

1. Thrombocytosis >400,000/cc
2. Leukocytosis >12,000/cc (in the absence of fever or infection)
3. Leukocyte alkaline phosphatase >100 IUM
4. Serum vitamin B_{12} >900 pg/ml

Polycythemia

Physiology	Etiology	Laboratory
1. Spurious polycythemia (contracted plasma volume)	Fluid loss (GI, renal, skin) Chronic stress Smoking	Normal red cell mass Significant orthostatic BP and pulse changes
2. True polycythemia (increased red cell mass) a. Marrow response to hypoxia	Acidosis Elevated altitude Smoking/COPD Carboxy- or met-hemoglobin	↓ pH Elevated carboxy- or methemoglobin Hb rises 1 g/dl for every 3–4% fall in O_2 saturation
b. Excessive erythropoietin production	Renal artery stenosis Functioning renal tumor	↑ Serum erythropoietin Diagnosis by IVP, arteriogram
c. Autonomous erythrocytosis	Polycythemia vera	↓ Serum erythropoietin

Untreated, 15–20% of affected persons develop marrow fibrosis with extramedullary hematopoiesis. This may lead to thrombocytopenia, leukopenia, massive splenomegaly, and anemia. This syndrome is refractory to treatment and carries a prognosis of 2–3 years. Rarely, patients (<1%) will develop extremely resistant AML; the incidence of acute leukemia is increased to 4% following radioactive phosphorus treatment and to 10% after alkylating agent therapy.

Polycythemia Vera: Signs and Symptoms

Symptoms	Signs
Headaches	Ruddy, plethoric complexion
Bleeding/bruising	Splenomegaly (>70% at presentation)
Sweating	Thrombocytosis or leukocytosis
Pruritis	(≤50% at presentation)
Gout	

The goal of treatment is to correct symptoms resulting from a high circulating red cell mass and to prevent the high incidence of thrombotic/bleeding episodes. Phlebotomy should be instituted at a rate of 300–500 cc two to three times a week to stabilize the hematocrit in the 45–47 vol% range. The relative contributions of thrombocytosis, platelet dysfunction, and increased red cell mass leading to hyperviscosity and thromboembolic complications is uncertain. Reducing the platelet count under 1,000,000/cc and treating erythrocytosis with phlebotomy are usually successful in alleviating bleeding and thrombotic complications. Frequent phlebotomy can lead to severe iron deficiency; although mild iron deficiency is helpful in suppressing erythrocytosis, severe deficiency may be symptomatic and further promote thrombocytosis. Most complain of pruritis; cholestyramine and cyproheptadine have been effective in treating this condition. Symptomatic gout may also develop and require treatment with allopurinol.

The indications for myelosuppressive therapy include thrombocytosis >1,000,000/cc (or 750,000–1,000,000/cc with thrombotic symptoms), phlebotomy requirements more frequently than every 6–8 weeks, or failure to control hypermetabolic symptoms. ^{32}P is administered in a single dose with effects occurring over a 2- to 3-month period of time. A remission often is seen, lasting several months to years. This treatment is preferred in frail elderly. Alkylating agents including busulfan, chlorambucil, and cyclophosphamide have also been suggested; these regimens, however, have severe side effects (pulmonary fibrosis, high leukemic risk, and hemorrhagic cystitis, respectively) and may lead to overtreatment thrombocytopenia or neutropenia.

IX. DISORDERS OF HEMOSTASIS

Platelet disorders include an absolute decrease or increased in platelet number and platelet dysfunction. The normal range for platelets is $150–400 \times 10^6/cc$, unchanged in the elderly. Bleeding is unusual with platelet counts $>50 \times 10^6/cc$. Although unclear, spontaneous hemorrhage may also complicate thrombocytosis, particularly with levels $>1,000,000/cc$. Platelet function may only intermittently be abnormal. Platelet function is best determined by measuring the bleeding time, with normal values ranging between 3 and 8 min.

Thrombocytopenia results from a disturbance of platelet production, distribution, or destruction. The evaluation must include examination of the peripheral smear, cessation of all medications if possible, and marrow examination to evaluate production abnormalities.

Thrombocytopenia: Mechanisms

Inadequate production
1. Marrow hypoproliferation: Aplastic anemia, drug reaction (thiazides, ethanol, estrogens), infections, myelosuppressive chemotherapy
2. Marrow replacement: Tumor, infection, sarcoid
3. Marrow abnormalities: Inherited and acquired genetic disorders (thrombasthenia, ADP storage disease)
4. Ineffective thrombopoiesis: B_{12}/folate deficiency, preleukemia, diGuglielmo's syndrome

---

(continued)

Thrombocytopenia: Mechanisms (*Continued*)

Abnormal distribution
1. Splenomegaly with sequestration
Increased destruction
1. Immunologic: ITP, drug reaction (quinine, digoxin, thiazides, estrogens, ethanol), lymphoproliferative disorders (lymphoma, CLL), SLE, infectious mononucleosis, frequent recipient of blood transfusions contaminated by platelets
2. Consumptive
   a. Activation of intravascular coagulation (DIC from surgery, ob/gyn complications, malignancy, gramnegative bacteremia, hemolysis, anaphylaxsis, snakebite, burns, SLE)
   b. Adherence to sites of endothelial damage: Surgery, trauma, vasculitis, prosthetic valves, TTP, hemolytic uremia syndrome

Idiopathic thrombocytopenic purpura (ITP) is a disorder usually affecting younger adults. However, secondary chronic ITP results from an autoimmune antiplatelet antibody and may present at any age with symptoms of thrombocytopenia without splenomegaly. Major concerns are spontaneous hemorrhage, especially intracranial bleeding. Approximately 50% of patients will respond to 1 mg/kg of prednisone daily, with an additional 10–20% responding at a 2 mg/kg dosage. If remission is not induced within 6–8 weeks, however, it is wise to consider splenectomy, thus avoiding side effects of long-term steroid administration.

Splenectomy provides a sustained remission in 60–70% of patients and a partial remission in an additional 10–20% of cases. Patients achieving a remission typically have postoperative platelet counts exceeding 500,000/cc. A surgeon with experience in testing patients with ITP should be consulted. Relapse may be seen in a small fraction of patients, resulting from hypertrophy of a small accessory spleen not apparent at surgery; CT or liver spleen scan can often demonstrate this finding. Persistant ITP may respond to vincristine therapy. (See table on p. 228.)

Consumptive platelet destruction reflects activation of the intravascular coagulation cascade and therefore is characterized by depression of individual clotting factors as well as levels of fibrinogen.

Chronic ITP*

Patient characteristics	Symptoms	Signs	Laboratory findings
Idiopathic Age 20–40 ♀ : ♂ = 3 : 1	Petechiae Bleeding Bruising	No lymphadenopathy No splenomegaly If idiopathic, no evidence of a primary disease	Platelets 20–100,000/cc Bleeding time >10 min Normal PMNs and RBCs
Secondary Age >40 with a primary disease	Intracranial bleeding		Antiplatelet antibody

*Normal marrow exam with adequate or increased megakaryocytes.

Thrombocytopenia: Laboratory Findings

Thrombocytopenia
Hypofibrinogenemia
PT prolongation
Increased fibrin degradation products
Decreased factor V and VIII levels
Microangiopathic blood film

Severe bleeding should be supported as needed with platelet concentrates and if necessary cryoprecipitates capable of providing fibrinogen and clotting factors. Treatment should focus on the underlying condition, and only in cases of severe unresponsive bleeding should heparin therapy be started at 10,000–20,000 units/day.

In the case of thrombocytopenia in the setting of aplastic anemia and absence of bleeding or surgery, platelet transfusions should be avoided, thus preventing development of platelet alloantibodies. Washed platelets and single-platelet-donor packs may prevent this complication. A platelet count performed 1 hr after transfusion will determine the maximal rise from platelet therapy: maximal life span for these cells is 4 days or less.

Thrombocytosis (platelet count >1,000,000/cc) may be autonomous or reactive. Treatment of autonomous thrombocytosis follows protocols similar to that used for polycythemia vera and CML with ^{32}P, busulfan, or other alkylating agents. (See table on p. 230.)

Problems in coagulation are typically inherited metabolic disorders affecting either plasma procoagulants (classic hemophilia, dysfibrinogenemia) or their inhibitors (antithrombin III). These rarely present in the geriatric age group. Antithrombin III deficiency (heparin cofactor values 40% of normal), however, may be acquired during later life. This disorder presents with resistance to heparin and recurrent episodes of venous thrombosis or pulmonary emboli. This is most commonly associated with liver disease and results from a hypoproduction of antithrombin III. Other etiologies include nephrotic syndrome, diabetes, L-asparaginase therapy, mixed oral contraceptive usage, disseminated intravascular coagulopathy, or diffuse atherosclerosis.

Thrombocytosis

Type	Etiology	Diagnosis	Prognosis
Reactive	Iron deficiency Infection Trauma Solid tumors	Marrow megakaryocytes suppressed	Rare symptoms despite platelet counts to 1,500,000/cc Resolves after condition treated
Autonomous	Idiopathic Stem cell disease: polycythemia vera, CML, myelofibrosis	Elevated marrow megakaryocytes Abnormal platelet aggregation to epinephrine	Bleeding or thrombotic episodes common with platelet counts >1,000,000/cc

X. DISORDERS OF LEUKOCYTES

Problems with granulocytes in the elderly typically encompass disorders of number rather than function. Most functional difficulties are congenital and present earlier in life with recurrent infections; granulocyte function is not normally affected by age. Although the heterozygous form of G6PD deficiency may lead to a decrement in hydrogen peroxide (H_2O_2) production and bacterial killing in leukocytes, this does not typically present a clinical problem. Toxic-granulation neutrophils during episodes of acute infection do show a decrease in chemotaxis, but this is reversible and not age related.

Granulocytosis is functionally defined as a polymorphonuclear cell count over 8000/cc. Often a granulocytic reaction in the elderly will include release of premature forms without an increase in absolute granulocyte count. Etiologic causes are numerous. Demargination resulting from emotional stress or dehydration is rarely of significant duration and should not postpone further evaluation for leukocytosis in the elderly.

Numerical Granulocyte Defects

Granulocytosis (PMN >8000)

1. Physiological: exercise, hemorrhage
2. Infection
3. Demargination: burns, endotoxin, electrical shock, drugs (e.g., epinephrine)
4. Neoplasia: bronchogenic, gastric
5. Metabolic: ketoacidosis, Cushing's disease
6. Hematologic malignancy: AML, polycythemia vera
7. Inflammatory: vasculitis, colitis

Granulocytopenia (PMN <2000)

1. Stem cell hypoproduction: aplastic anemia
2. Maturational arrest in marrow
   a. Cyclic: regularly decreased each 21–35 days (reversible arrest)
   b. Familial: normal stem cells (probable chronic arrest)
   c. Megaloblastic: EtOH, folate deficiency
   d. Marrow invasion: neoplasia

(*continued*)

Numerical Granulocyte Defects (Continued)

Granulocytopenia (PMN <2000) (Continued)

3. Immune destruction: antigranulocyte IgG or IgM antibody; or T-cell suppression of CFU-G factors from monocytes; in Felty's, lupus, drug reaction (chloramphenicol, Au+, chemotherapy)
4. Nonimmune destruction: hypersplenism

Leukemoid reactions can easily be confused with laboratory features of leukemia. Typically the total cell count is elevated and displays immature forms not normally found in the peripheral circulation. Diagnostic tests including marrow aspiration and biopsy may be of assistance in ruling out metastatic disease. Karyotype analysis for Ph_1 (Philadelphia) chromosomes and leukocyte alkaline phosphatase helps rule out CML.

Appropriate infectious disease studies are also warranted. In many cases, the diagnosis may remain unclear until spontaneous resolution of the leukemoid reaction occurs as the stress event resolves or is treated. (See table on p. 233.)

Granulocytopenia is classically defined as polymorphonuclear cell counts below 2000–2500 cells/cc. Severe, persistent neutropenia in the absence of a cyclical or familial history usually warrants bone marrow examination and a determination of antigranulocyte antibody.

Eosinophilia with total eosinophil counts over 700–750 develops in the elderly from similar causes as in the young. Allergic disorders, parasitic infections, malignancy (particularly Hodgkin's), hypereosinophilic pulmonary syndromes, dermatologic disorders (eczema, pemphigus), Addison's disease, and drug reactions are the most common causes.

XI. DISORDERS OF IMMUNE FUNCTION

T-lymphocyte function and subpopulations change significantly with increasing age. The loss of thymic activity in differentiation of immature T cells cannot be reversed by injections of young thymus extract. This leads to a deficiency in T-cell function, far more important than the decrease noted in cell number. In particular, the T-cell response to new antigens is poor. Screening of apparently

Leukemoid Reactions: Description and Etiology

Leukemoid type	Cell count/differential	Clinical disease setting
Myeloid	PMNs >50,000/cc with >2% myeloblasts or myelocytes Occasionally PMNs are 25,000–50,000/cc, primarily consisting of immature forms	Infections: TB, meningitis, pneumonia, sepsis, diphtheria, glomerulonephritis Malignancy: gastric, Hodgkin's, breast, lymphoma, adrenal, lung Toxins: mercury, eclampsia, bun Marrow stimulants: acute hemorrhage, recovery from myelosuppression Inflammatory: RA, dermatitis herpetiformis
Lymphoid	Lymphocytes >40% with immature forms and total leukocyte count elevation	Infections: mononucleosis, TB, CMV, pertussis, varicella, mumps Malignancy: gastric, lymphoma, breast, melanoma Inflammatory: exfoliative dermatitis
Monocytoid	Monocytes >30% and total leukocyte count >30,000	Infections: TB, congenital syphilis Malignancy: teratoma Infection/marrow stimulant: cholecystitis in a patient with myelodysplasia

healthy elderly for delayed cutaneous hypersensitivity shows poor response to new antigens (DNCB); residual response is noted, however, in most elderly to at least one familiar antigen in a battery of six. This is particularly true if they are rechallenged 10–14 days later to enhance amnestic responses. This is in contrast to a normal of three to four responses to a battery of six in healthy young adults.

Changes in Immune Function in the Elderly

Cellular immunity
    Thymic atrophy: 5–10% function remaining at age 50
        ↓ T suppressor and/or killer cell numbers
        ↓ Proliferation in anti-Dr autologous mixed lymphocyte
        reactions
        ↓ Proliferation to new antigens

Humoral immunity
    Total numbers of B cells unchanged
    ↓ Antibody production to new or old antigens:
    probable T-helper-cell defect
    ↑ Autoantibody production:
    probable T-suppressor-cell defect

Total B-cell numbers are usually unchanged in the humoral system; functional parameters, however, are modified with age. Antibody production in response to a specific stimulus is decreased; paradoxically, autoantibody production increases. This suggests a loss of T-cell regulation rather than an intrinsic B-cell deficit. Aging individuals develop autoantibodies without clinical symptoms in most cases. Common autoantibodies that develop in the elderly include anti-parietal-cell, antimitochondrial, antithyroglobulin, anti-nucleic-acid (d-DNA and s-DNA), anti-smooth-muscle, and immunoglobulin (anti-Fc portion of IgM or rheumatoid factor).

Loss of regulation for a specific antigen, leading to benign monoclonal gammopathy, increases in incidence to 3% over age 70. Patients with this problem usually have a monoclonal spike of less than 3.0 g/dl (for IgG) or less than 1.5 g/dl (for IgM or IgA) and no lesions suggesting malignancy: negative Bence Jones protein, absent bone lesions, absent hepatosplenomegaly, and <10% plasma cells on marrow examination. In 10% of cases, however, there is a progression to Waldenstrom's, myeloma, or amyloid. No clear-cut prognostic factor has been identified. Patients who continue to have asymptomatic hyperplasia of the active B-cell clone must be followed regularly for signs of spike increase or other lesions. Treatment of associated inflammatory, infectious, or hepatic diseases may occasionally lead to remission of the gammopathy.

SUGGESTED READING

Ahn Y. S., Harrington W. J., Mylragenam R., et al: Danazol therapy for autoimmune hemolytic anemia. *Ann Intern Med* 102:298–301, 1985.

Dallman P. R., Yip R., Johnson C.: Prevalence and causes of anemia in the United States 1976–1980 (NHANES II). *Am J Clin Nutr* 39:437–445, 1984.

Delafuente J.: Immunosenescence. *Med Clin North Am* 69(3):475–486, 1985.

Evans D. L., Edelsohn G. A., Golden R. A.: Organic psychosis without anemia or spinal cord symptoms in patients with vitamin B_{12} deficiency. *Am J Psychiatry* 140:218–221, 1983.

Foucar K., Langdon R. M., Armitage J. O., et al: Myelodysplastic syndromes: A clinical and pathologic analysis of 109 causes. *Cancer* 56:553–561, 1985.

Hansen N. E.: The anemia of chronic disorders: A bag of unsolved problems. *Scand J Haematol* 31:397–402, 1983.

Lerner W., Caruso R., Faig D., et al: Drug dependent and nondrug dependent antiplatelet antibody in drug induced immunologic thrombocytopenic purpura. *Blood* 66(2):306–311, 1985.

Levine D. S.: Intestinal vascular ectasia: Improving diagnostic capability poses therapeutic dilemma. *Am J Med* 76:1151–1155, 1984.

Szklo M., Sensenbrenner L., Markowitz J., et al: Incidence of aplastic anemia in metropolitan Baltimore: A population based study. *Blood* 66(1):115–119, 1985.

10

Oncologic Problems in the Elderly

I. INTRODUCTION

With the exception of childhood leukemias and central nervous system tumors, age is the single greatest risk factor for developing a malignancy. Sixty percent of all cancer occurs after age 60 with an incidence reaching almost one in ten by age 70. Possible etiologic factors that have been suggested include the cumulative exposure to carcinogens (both initiators and promoters) or altered host regulation mechanisms resulting in decreased immune surveillance. Disorders of immune function predispose to increase rates of malignancy. This can be demonstrated in patients with hereditary immunodeficiency, organ transplantation with immunosuppression, autoimmune disease, and following curative chemotherapy for other malignancies. Altered regulation with aging may lead to activation or "derepression" of oncogenes already present in the genome. Alternatively, the deficit may reflect a decreased ability to repair somatic mutations in DNA.

At present, there is an increasing incidence of lung, pancreatic, and prostatic cancer rates. In contrast, gastric, hepatic, and uterine cancer rates are falling.

1982 Cancer Incidence: All Ages		
Malignancy	Men	Women
Lung	22%	9%
Colorectal	14%	15%
Breast	—	26%
Prostate	18%	—
Uterus	—	13%
Leukemia/lymphoma	8%	7%
Pancreas	3%	3%
Skin	2%	2%
Oral	6%	2%
Urinary	9%	4%
Ovary	—	4%
Other	18%	15%

II. PARANEOPLASTIC SYNDROMES

Paraneoplastic syndromes are defined as distant or remote effects of a neoplasm, unrelated to the site of primary or metastatic tumor involvement. These syndromes are thought to be caused by specific mediators or products of derepressed genes. These products, normally excluded by intact cellular basement membranes, are now released into the peripheral circulation. Other possible causes include autoimmunity (possibly related to or distinct from the previous factors) and loss of inhibitory regulation. (See table on pp. 240–241.)

The syndromes have provided markers for initial diagnosis, follow-up for tumor relapse, and understanding of the underlying cell type. Treatment of these syndromes may provide significant improvement not only in the quality but also in the quantity of remaining life.

III. THE LEUKEMIAS

Myelogenous or nonlymphocytic leukemia occurs as a result of a malignant transformation in a pluripotent stem cell. Acute myelogenous leukemia (AML) accounts for 80% of acute leukemias in adults with a peak incidence of 15/100,000 after age 70. The clonal etiology is confirmed by common karyotypic translocations in all cells and one single G6PD allele in heterozygous women with AML. Typical signs, symptoms, and laboratory findings are listed.

Acute Nonlymphocytic Leukemia: Presentation

Symptoms/Signs	Laboratory Findings
Fatigue, pallor	Anemia
Bruising, bleeding	Thrombocytopenia
Fever, infections	Neutropenia
Anorexia	Karyotype rearrangements
Splenomegaly	Hypercellular marrow with
Leukostastis: hypoxia,	30% blasts
mental changes, anuria	15% blast cells in peripheral circulation

Untreated, most patients live less than 6 months. Most patients require treatment within days of diagnosis for bleeding or infection. Myelogenous leukemia can be divided into seven subtypes. All subtypes have similar presentation; prognostically, however, M-3 (progranulocytic) has a significant incidence of disseminated intravascular coagulapathy, and types M-4 and M-5 (monocytoid) require prophylactic central nervous system chemotherapy.

Acute Myelocytic Leukemia: Classification

Designation	Common Terminology	Predominant Cell Type
M-1	Undifferentiated	Granulocytes
M-2	Acute myelocytic	Granulocytes
M-3	Progranulocytic	Progranulocytes
M-4	Acute myelo-monocytic	Granulocytes + monocytes
M-5	Acute monocytic	Monocytes
M-6	Erythroleukemia	Granulocytes + erythrocytes
M-7	Acute megakaryo-blastic	Granulocytes + megakaryocytes

Paraneoplastic Syndromes

Syndrome	Mechanism	Tumors
Metabolic		
Hypercalcemia	Ectopic PTH or PTH-like substance	Squamous cell cancer of respiratory or foregut, lymphoma, ovary, colon, renal
	Lytic metastasis (prostaglandins activated)	Breast cancer, others
	Osteoclast-activating factor	Multiple myeloma
Hypoglycemia	Insulin production	Insulinoma, hepatoma, gastric, retroperitoneal sarcoma, adrenal
Hypertension	Renin or aldosterone production	Renal cell cancer, pheochromocytoma
Hyperuricemia	High cellular turnover rates	Leukemia, lymphoma, germ cell neoplasm
Hyponatremia	Antidiuretic hormone production	Small-cell cancer of lung, pancreas, other APUD tumors
Hyperglycemia, DKA, alkalosis	ACTH production	Small-cell cancer of lung, pancreas, thyroid, gastric, ovary, other APUD tumors
Rheumatologic and musculoskeletal		
Paraneoplastic carcinomatous polyarthritis	Unknown (MCPs, PIPs, other large joints, assymetric)	Adenocarcinoma

Condition	Etiology	Associated malignancy
Hypertrophic osteoarthropathy	Unknown (periosteal reaction with calcification)	Lung cancer
Sjogren's, systemic vasculitis, polymyositis	Autoantibodies	Adenocarcinoma, lymphoma
Neurological		
Myasthenia	Antibody to ACh receptor	Thymoma
Eaton–Lambert	Impaired ACh release, unkown etiology	Small-cell cancer of lung, adenocarcinoma
Hematologic		
Thromboemboli (PE, DVT)	Hyperfibrinogenemia, unknown etiology	Adenocarcinoma
Polycythemia	Erythropoietin production	Renal cell cancer, hepatoma, gastric, ovary, sarcoma
Autoimmune hemolytic anemia	Antierythrocyte antibody (Rh locus)	Hodgkin's, lymphoma, B-cell leukemia
Gastrointestinal		
Recurrent PUD	Gastrin production	Islet cell tumors
Dermatologic		
Acanthosis nigricans	Unknown	Adenocarcinoma
Vitiligo	Unknown	Melanoma, gastric
Exfoliative erythroderma	Unknown	Lymphoma

Patients who present with severe leukostasis (blast counts
>100,000/dl) require prompt treatment with leukapheresis or hy-
droxyurea to prevent hemorrhage and thrombosis. Induction chemotherapy combines cytosine arabinoside (Ara-C) and daunorubicin.
Typically up to 70% achieve a complete remission. Although
consolidation regimens are still controversial, most patients will
relapse within a few months without it. Regimens include Ara-C,
daunorubicin, and other alkylating agents, usually maintaining a
remission of 10 to 20 months. Twenty percent will achieve a
remission lasting 5 years. Bone marrow transplantation during first
remission is usually reserved for those under 45 years of age.
Elderly patients have no noticeable increase in deaths during
induction-phase aplasia; in addition, remission rates appear not to
be affected by age. Unfortunately, nonspecific symptomatology
often delays the diagnosis in the elderly, a problem leading to
increased leukostasis and early deaths.

Chronic myelogenous leukemia (CML) represents 15 to 20%
of all leukemia in adults. The peak incidence occurs approximately
at age 40, with a range from 25 to 60 years. The mutation involves
a pluripotent stem cell leading to granulocytes, erythrocytes, and
megakaryocytes. The typical genetic translocation (Philadelphia
chromosome) does not affect granulocyte function or lead to an
increased incidence of infections. Typical signs, symptoms, and
laboratory findings are listed.

Chronic Myelogenous Leukemia: Presentation

Signs/Symptoms	Laboratory Findings
Anorexia	Splenomegaly (90%)
Fatigue	Leukocytosis (WBC
Abdominal fullness	>20,000); some imma-
Bone pain	turity (blasts 5–10%)
Advanced disease: retinal	Hypercellular marrow;
hemorrhages, bruising,	some fibrosis
chloromas	Philadelphia chromosome
	(Ph') positive (90%); ↓
	leukocyte alkaline phos-
	patase; ↑ serum LDH
	↑ Serum uric acid
	Advanced disease: anemia,
	thrombocytopenia

Untreated, there is a 5–10% mortality rate during the first year
of illness; this increases to 25% each year thereafter. Chronic-phase
treatment is currently designed to alleviate symptoms without
significant prolongation in life expectancy. Busulfan therapy has a
90% response rate; daily treatment for 1 to 3 months is required to
bring the white count to less than 20,000/mm^3. At this point,
busulfan should be discontinued to prevent marrow aplasia and
withheld until white counts rise to 50,000/mm^3 or greater with
symptoms. Typically, CML terminates in a myeloid or lymphocytic
blast phase, frequently heralded by busulfan resistance and new
chromosome abnormalities in the accelerated phase. Treatment of
Ph'-positive AML developing from CML is typically very difficult;
there is only a 10% response to Ara-C/daunorubicin regimens, and
relapse occurs typically within months. Lymphoid transformation is
more responsive despite Ph' remaining positive, with a 50% remission rate to vincristine/prednisone regimens. Unfortunately, relapse
usually occurs within 6 months. Allogeneic marrow transplantation
has been attempted during the chronic phase of illness; at the
current time, this form of therapy is reserved for patients under 45
years of age.

Lymphocytic malignancies present in a spectrum ranging from
acute lymphoblastic leukemia, representing a malignant transformation of a committed B-lymphocyte stem cell, to multiple myeloma,
involving differentiated and functioning plasma cells. The characteristics of each disease, including localization, rapidity of growth,
and clinical signs, therefore vary greatly and reflect the level of
differentiation achieved by the cell prior to its malignant transformation. Some types of T-lymphocytic lymphoma and T-cell acute
lymphocytic leukemia (ALL) have been associated with HTLV-I
(human T-lymphocyte virus). It is also suspected that retroviruses
play a significant role in malignant B-cell transformation. Epstein–
Barr virus is linked etiologically to both Burkitt's lymphoma and
nasopharyngeal carcinoma.

Acute lymphocytic leukemia (ALL) is primarily a disease of
children but does have a second peak during young adulthood. The
incidence of Ph'-positive ALL rises to 10% in adults, suggesting
lymphocytic blast transformation of a preexisting CML. As noted
earlier, blast transformation to Ph'-positive ALL may be treated but
rarely remains in remission longer than 6 months. Classification by
lymphocyte type transformed is listed. Most adult cases are L-2
type, which encompasses several cell types; immunologically, 50%

are pre-B, 20–40% null cell, 10–20% T-cell, and rarely a B-cell
type can be found.

Classification Scheme for Acute Lymphoblastic Leukemia

French–American–British System	Immunologic Criteria
L-1: Small cells, scant cytoplasm, rare nucleoli	T-cell: Sheep red blood cell rosetting present; terminal deoxytransferase (TDT) positive; T-antigens on surface
L-2: Large cells, clefted nuclei, one to several nucleoli	B-cell: Surface immunoglobulin positive; Ia antigen positive; TDT negative; common ALL antigen (cALLa) negative
L-3: Large cells, vacuolated nuclei, many nucleoli	Common premature B-cell (pre-B): cALLa positive; TDT positive; cytoplasmic μ chains present
	Null cell: cALLa negative; surface immunoglobulin negative; T-antigen negative

---

Few centers have extensive experience treating ALL in adults;
existing protocols consist of vincristine, prednisone, and doxorubicin or daunomycin, with possible additions of L-asparaginase
or cytoxan. Intrathecal methotrexate is mandatory for CNS prophylaxis. Poor prognostic factors include male sex, advanced age,
CNS disease present at onset, and chromosonal abnormalities.
Using these protocols, the remission rate for adults over age 30
falls to 60–70% (as compared with an 80–90% rate reported in
children); maintenance chemotherapy is mandatory for at least 3
years because of frequent relapses. Approximately 20% of adult
patients will remain disease-free permanently. Although marrow
transplantation is being used primarily during the first remission,
only patients under 45 years of age are currently considered for this
form of therapy.

Chronic lymphocytic leukemia (CLL) is the most common form of leukemia in the United States, comprising 30% of the total. Nearly all cases represent the clonal proliferation of a single mature B-cell, typically synthesizing μ heavy chains (with or without concomitant δ chains). There is a 2:1 male predominance, and cases rarely occur before age 50; median age at onset is 60. The malignant lymphocytes carry μ chains and Ia and Ig Fc receptors as cell surface markers. Up to 25% of patients are asymptomatic at time of presentation; typical signs and symptoms for symptomatic patients are listed.

Chronic Lymphocytic Leukemia: Presentation

Signs/Symptoms	Laboratory Findings
Painless adenopathy	Lymphocytosis
Fatigue, pallor	>15,000/mm^3
Fever, pyogenic infections	Anemia (normochromic,
Anorexia	normocytic)
Splenomegaly	Hypogammaglobulinemia (50%)
	Monoclonal spike (3–5%)

Patients rarely will present with significant marrow displacement and organomegaly. In these cases, there is usually a rapid progression to death within 2 years. In most cases, the disease is indolent, and the indications for initiating treatment are unclear. Clinical staging and prognosis are listed.

Staging for Chronic Lymphocytic Leukemia

Stage	Clinical stage	Survival
0	Lymphocytosis 15,000/mm^3	10 years
1	Lymphocytosis plus lymphadenopathy	8 years
2	Lymphocytosis plus lymphadenopathy and hepatomegaly or splenomegaly	7 years
3	Lymphocytosis plus lymphadenopathy and anemia (hemoglobin 11 g/dl)	2–5 years
4	Lymphocytosis plus lymphadenopathy and thrombocytopenia (platelets 100,000/mm^3)	2 years

Local lymph node enlargement or splenomegaly causing symptoms may be treated with irradiation; immune-mediated hemolytic anemia, thrombocytopenia, or neutropenia is best controlled with steroid administration. Since no increase in survival has been shown by initiating immunosuppressive treatment during assymptomatic stages (I and II), systemic therapy is best delayed until organomegaly or cytopenias become symptomatic. Single agents of choice are chlorambucil (0.1–0.2 mg/kg per day) or cyclophosphamide (2–3 mg/kg per day); prednisone can also be used concomitantly at doses of 30–60 mg/m² surface area to treat immune phenomena as well as to decrease lymphadenopathy rapidly. Patients are treated intermittently, allowing 2 to 4 weeks for marrow recovery. Terminal stages of the illness are associated with refractory anemia, thrombocytopenia, and rarely transformation to histiocytic lymphoma or leukemia. A future avenue of treatment may include the use of monoclonal antibodies directed toward the CLL idiotype specific for the individual patient.

Hairy cell leukemia is typically a disease of males aged 30–55, representing malignant transformation of mature B-cells. It is associated with HTLV-II infection; the pathognomonic finding is monocytoid cells with cytoplasmic spicule projections. Prominent early features include anemia and splenomegaly; later, leukopenia, bleeding diathesis, malaise, and infections are noted. There is little response to treatment; splenectomy can alleviate refractory cytopenias. Fifty percent of patients will survive 4 years following presentation of illness; the most common terminal event is an infection.

IV. MYELOMA AND GAMMOPATHY

Plasma cell myeloma is a malignant clonal disease of plasma cells, typically producing IgG heavy chains (rarely IgD or IgA) and either κ or λ light chains. In 10% of cases only light chains are produced. The disease rarely affects anyone below the age of 40; the mean age at time of diagnosis is 62. Peak incidence occurs at age 75, when it approximates 25/100,000. Blacks are affected more commonly than whites. The most common presenting symptom is back or chest pain. Other symptoms and complications are listed.

Multiple Myeloma: Complications

Renal	Hyperuricemia
	Hypercalciuria
	Light chain excretion may lead to
	Tubular dysfunction
	Acidosis
	Proteinuria
	Aminoaciduria
Hematologic	Hyperviscosity (blurred vision, epistaxis, headache)
	Anemia (pallor, fatigue)
	Amyloidosis
	Thrombocytopenia/ neutropenia (10–15% incidence)
	Cryoglobulins (cold urticaria)
	Leukemia
	Hypogammaglobulinemia: susceptibility to virus (herpes, varicella/zoster) and encapsulated bacteria [pneumococci (pneumonia) and gram-negative bacteria (pyelonephritis, septicemia, meningitis)]
Neurological	Spinal root compression
	Myelomatous meningitis
Dermatologic	Pyoderma gangrenosum

Tumor burden can be estimated from the size of the "M" spike and degree of anemia. Anemia along with renal function and hypercalcemia from lytic lesions are the most significant predictors of survival. Fifty to sixty percent of all myeloma patients at any state respond to chemotherapy; nonresponders have the poorest prognosis of all groups.

Multiple Myeloma: Diagnostic Criteria*

Major Criteria

I. Plasmacytomas on tissue biopsy
II. Bone marrow plasmacytosis >30% plasma cells
III. Monoclonal globulin spike with IgG >3.5 mg/dl or IgA >2.0 mg/dl or light chain excretion >1 g/24 hr

Minor Criteria

a. Marrow plasmacytosis 10–30%
b. Monoclonal globulin spike with IgG <3.5 mg/dl or IgA <1 mg/dl
c. Lytic bone lesions
d. Reduction in normal immunoglobulins: IgM <50 mg/dl, IgA <100 mg/dl, or IgG <600 mg/dl

Diagnosis Supported by

1. I or II plus b
2. I or II plus c
3. I or II plus d
4. III
5. a + b + c
6. a + b + d

*Developed by Southwest Oncology Group.

The median survival in patients with low tumor burden is 3 years; this can be increased another 12–18 months with response to alkylating agent (melphalan) and prednisone chemotherapy. Multiagent i.v. regimens, including monthly carmustine, cyclophosphamide, and prednisone, have been compared to the oral regimen in the elderly. This form of treatment results in a 70–80% response rate compared to the 50–70% rate with melphalan alone. Patients developing resistance to melphalan may respond to other alkylating agents. No additional hematologic toxicity has been reported with the multiagent i.v. regimen when used in the elderly; gastrointestinal tolerance also appears to be better.

Terminally, the clinical course may be complicated by acute leukemia (<5%), persistent cytopenias (30%), renal failure, infection (45%), or other medical diseases (20%). Adjunctive treatment modalities are listed.

Multiple Myeloma: Durie and Salmon Staging

Stage	Myeloma cell mass	Criteria
I	Low	All of the following:
		Hemoglobin >10 g/dl
		Normal serum calcium
		Solitary plasmacytoma or normal bone
		Low monoclonal protein production:
		IgG >5 g/dl
		IgA >3 g/dl
		Urinary κ or λ <4 g/24 hr
II	Intermediate	Fits neither category I nor III
III	High	One or more of the following:
		Hemoglobin >8.5 g/dl
		Serum calcium >12 mg/dl
		More than three lytic bone lesions
		High monoclonal protein production:
		IgG >7 g/dl
		IgA >5 g/dl
		Urinary κ or λ >12 g/24 hr

Multiple Myeloma: Adjunctive Treatment

Prevention of renal disease
  Fluid intake (2000–3000 cc/day)
  Avoidance of IVP dye
Treatment of bone disease
  Activity, physical therapy
  Analgesics, orthopedic supports
  Androgens, vitamin D, adequate calcium
  Radiation therapy to painful areas (local)
Treatment of anemia
  Androgens
Treatment of hypercalcemia
  Furosemide-induced diuresis
  Calcitonin, 100–200 units s.c. q12h
  Mithromycin

Macroglobulinemia, characterized by an increase in serum IgM, exists in both a clonal and polyclonal variety. Approximately one-third of cases are benign (cold agglutinin disease, IgM benign

monoclonal gammopathy); the rest involve B-lymphocytic neoplasms (Waldenstrom's, IgM myeloma or plasmacytoma, diffuse
lymphoma, or CLL). Typically, Waldenstrom's macroglobulinemia
results in the highest elevation of monoclonal proteins. This disease
has a mean age at onset of 64. Common presenting findings include
fatigue, bleeding diathesis, weight loss, and neurological complaints. Common physical and laboratory findings are listed.

Waldenstrom's Macroglobulinemia: Signs

Physical Findings	Laboratory Findings
Hepatomegaly (1/3)	Anemia (normochronic,
Splenomegaly (1/3)	normocytic)
Lymphadenopathy (1/3)	↑ Serum viscosity
Ophthalmologic (1/3)	(normal: serum/H_2O =
(see hyperviscosity)	1.5)
Purpura, dependent (1/5)	Prolonged bleeding time
Neurological (1/5)	Bence Jones proteinuria
(see hyperviscosity)	(1/4)
Congestive heart failure	Nonspecific proteinuria
(rare)	(1/10)
Symptoms of	Uremia (1/4)
cryoglobulinemia:	Cryoglobulinemia (1/3)
Reynaud's phenomenon	Immunoelectrophoresis
Cold urticaria	spike
	Osteolytic bone lesions
	(1/50)

Bone marrow biopsy reveals diffuse hyperplasia of lympho-
cytoid cells and immunoblasts with a slight increase in plasma
cells.

Differentiating myeloma from Waldenstrom's macroglobulinemia is possible by noting the absence of bone lesions and the
presence of adenopathy/hepatosplenomegaly. Although renal disease occurs in both, it is typically a glomerular defect caused by
IgM deposition rather than the tubular dysfunction caused by light
chain toxicity seen in myeloma. The primary etiology for bleeding
is platelet dysfunction. A "hyperviscosity syndrome" may develop
when serum viscosity is increased to four to five times that of

water; nearly all patients will have symptoms when serum viscosity exceeds tenfold that of water. Improvement can usually be noted following treatment customarily given for myeloma. Management of severe anemia or neurological compromise may need to include acute plasmapheresis for removal of macroblobulin-rich plasma.

Hyperviscosity Syndrome: Signs and Symptoms

Ocular
   Disturbance of vision
   Retinal hemorrhage
   "String of sausages" (beading) appearance of retinal
     veins
Hematologic
   Slow bleeding: nasal, gingival, GI, purpura
   Prolonged bleeding with minor surgery
   Anemia, fatigue
Neurological
   Headache, vertigo, nystagmus
   Postural hypertension, dizziness
   Hearing loss
   Polyneuropathy
   Lethargy, somnolence, stupor
   Seizures
Cardiovascular
   Congestive heart failure resulting from expanded plas-
     ma volume
Systemic
   Anorexia
Renal
   Concentration defects
   Glomerular deposits of IgM or amyloid

Prognosis varies considerably. In fact, some patients free of hepatosplenomegaly have been reported to have a very slowly progressive course lasting over 8–10 years. Treatment should be initiated when anemia or neurological complications become symptomatic. Aklylating agents (chlorambucil, cytoxan, melphalan) are typically used, with increases in survival as much as 1–2 years in those who respond.

V. THE LYMPHOMAS

Hodgkin's disease is a malignant disease of the lymphoid
tissue characterized by classic Reed–Sternberg cells in a reactive
milieu of lymphocytes, eosinophils, and fibroblasts. Reed–
Sternberg cells are histiologically of macrophage–histiocytic lineage
and represent a failure of cellular division following nuclear replica-
tion. The disease has an atypical bimodal incidence, with an early
peak at 26–28 (5/100,000) and a slow rise after age 40, peaking at
age 75 (6/100,000). Epidemiologic studies of those affected earlier
in life correlate with lower socioeconomic status, suggesting an
infection with a virus of low oncogenic potential. Although T-cell
immunity is depressed, humoral immunity is maintained until
advanced stages. Typically, the disease presents with adenopathy
progressing to involve contiguous lymph node areas. Later in the
disease, particularly following splenic involvement, distant sites
may become involved through hematogenous dissemination. Prognosis depends on clinical stage and pathological subtype. Older
adults typically present with a mixed-cellularity subtype and a
higher stage at presentation (40–50% stage III or IV; 40–50% with
"B" symptoms), often complicated by fever of unknown origin,
jaundice, or cytopenias.

Treatment protocols differ radically according to the clinical
stage of the disease. For this reason, complete clinical staging is
mandatory. For clinical stages I–III, the traditionally accepted
evaluation includes a detailed physical examination, chest radiograph, bone marrow biopsy, lymphangiogram, and staging
laparotomy. Recently, the less invasive abdominal CT scan has
replaced laparotomy in some cases. A 20–30% false-positive and
-negative rate has been noted, particularly with regard to splenic
involvement. In cases in which differentiating stage II from stage
III disease is of paramount importance, a laparotomy should still be
performed if otherwise clinically acceptable. Stage IV disease only
requires a positive liver, marrow, or other extralymphatic organ
biopsy.

Treatment of Hodgkin's disease in patients over age 60 is often
complicated by the presence of other underlying medical conditions.
These may preclude adequate staging or reduce tolerance for
chemotherapy and radiotherapy, leading to abbreviated and often
less curative courses. Adequately staged and treated elderly adults
in stages I to IIIa have 5-year freedom-from-relapse and survival
rates of 80% and 85%, respectively. Most deaths result from

Hodgkin's Disease: Ann Arbor Clinical Staging and Acceptable Treatment Intensity

Stage	Description of involvement	Adequate treatment
I	One lymph node region or extralymphatic origin/site	Subtotal lymphoid irradiation
II	Two or more lymph node regions on the same side of the diaphragm	A*: Total lymphoid irradiation B†: Total lymphoid irradiation
III	Lymph node regions involved on both sides of the diaphragm; additional involvement may include spleen and/or an extralymphatic organ/site	A*: Total lymphoid irradiation, with hepatic radiation of spleen +/OR combination chemotherapy (MOPP, ABVD) B†: Combination chemotherapy with or without irradiation
IV	Diffuse involvement of one or more extralymphatic organs or tissues; lymph node enlargement may or may not be present	Combination chemotherapy with or without irradiation

*A: lacking "B" symptoms.
†B: (1) unexplained fevers to 38°C; (2) night sweats; and (3) unexplained weight loss of 10% total body weight.

Hodgkin's Disease: Histological Subtype

Type	Initial presentation in clinical stage I or II	Frequency of presentation
Lymphocyte predominance	85%	10%
Nodular sclerosis	70%	40–70%
Mixed cellularity	50%	30–50%
Lymphocyte depletion	30%	5–10%

unrelated conditions and not from progressive Hodgkin's disease. Inadequately staged patients (no laparotomy) in stages I through IIIa have significantly reduced 5-year survival rates (30–35%). Elderly patients staged IIIb or IV rarely receive full chemotherapy doses. Of those who respond to treatment, life expectancy can be increased as much as 3–4 years, with relapses common thereafter.

Non-Hodgkin's lymphomas (NHL) originate from all types of lymphocytes, ranging from stem cells to activated immunoblasts and well-differentiated T cells. Rarely, the cell type may be truly histiocytic or tissue macrophage in origin. Therefore, the clinical characteristics and course of this disease vary considerably. The major predisposing factor for development of NHL appears to be loss of normal immune system functioning. The diseases showing a greatly increased risk of NHL include immune deficiencies (Wiskott–Aldrich syndrome, IgA deficiency, common variable immunodeficiency), autoimmune states (rheumatoid arthritis, Sjogren's syndrome, Cebac disease), and immunosuppressed states (particularly transplant recipients). A preexisting viral infection by EBV or RNA viruses may also be required for malignant transformation. On this basis, it is not surprising that the incidence of NHL rises steadily with age, reaching a peak of 50/100,000 at age 70.

Staging follows the Ann Arbor criteria for Hodgkin's disease. Diffuse disease must be distinguished from local disease occurring in one or two nodal groups. This latter form is potentially curable by radiotherapy alone. The traditionally used Rappaport criteria are being replaced by a working classification based on cell type and tumor grade. Low-grade tumors typically present in later stages, more commonly have an indolent course, and respond poorly to chemotherapeutic regimens. High-grade NHL may present in an earlier stage, tends to be aggressive, and responds well to chemotherapy, with a cure even possible.

Patients present with adenopathy and/or systemic symptoms. The most common low-grade NHL is nodular poorly differentiated lymphocytic lymphoma (PDLL), which carries a median survival of 7 years and is not considered curable by either radiotherapy or chemotherapy in stages III or IV. Local radiotherapy to symptomatic nodal areas or single-alkylating-agent therapy with prednisone may be palliative. Ongoing protocols are currently evaluating multiagent regimens for potential cure. Patients who are truly stage I or II after laparotomy may be treated with intensive extended field irratiation, as in the case of Hodgkin's disease.

Non-Hodgkin's Lymphoma: Classification

Working formulation	Rappaport	Cell surface markers*
Low grade		
1. Malignant lymphoma, small lymphocytic	Lymphocytic, well differentiated	B
2. Malignant lymphoma, follicular, small cleaved cell	Nodular, lymphocytic poorly differentiated	B
3. Malignant lymphoma, follicular, mixed small/large	Nodular, mixed histiocytic–lymphocytic	B
Medium grade		
1. Malignant lymphoma, follicular, predominantly large cell	Nodular histiocytic	B
2. Malignant lymphoma, diffuse, predominantly small cleaved cell	Diffuse, lymphocytic, poorly differentiated	B
3. Malignant lymphoma, diffuse, mixed small/large	Diffuse, mixed histiocytic–lymphocytic	B,T,N
4. Malignant lymphoma, diffuse, large cell	Diffuse histiocytic	B,N
High grade		
1. Malignant lymphoma, large cell, immunoblastic	Diffuse histiocytic	B,T,N
2. Malignant lymphoma, lymphoblastic	Lymphoblastic	T,N
3. Malignant lymphoma, small, noncleaved cell	Undifferentiated, Burkitt's	B

*B, surface IgG receptors; T, sheep erythrocyte receptors; N, neither.

Diffuse large-cell lymphoma is the most common medium-grade NHL, with a median survival of 1.5–2.5 years. Combination chemotherapy is standard treatment; alternate regimens usually include myelosuppressive and nonmyelosuppressive drugs with or without additional methotrexate. Treatment is carried to six courses or two courses past disappearance of disease; approximately 75% of those treated will respond, with 75% obtaining a long-term (>2-year) remission. Increased age (>60) appears to be a poor prognostic factor, as is the presence of anemia (Hb <10), "B" symptoms, GI involvement, and elevated serum LDH. Although high-grade malignancies rarely occur in the elderly, if present CNS preventative treatment is suggested.

VI. BREAST CANCER

The risk of developing breast cancer rises with age. In elderly
women, the disease largely occurs during the postmenopausal years.
The incidence rises with age to 314/1000 women over age 65,
peaking even higher at age 80. A full breast examination should
should be part of every elderly woman's physical examination
regardless of the reason for the visit.

Factors Suggested as Increasing the Risk of Breast Cancer in Elderly Women

Hyperestrogenemia
Nulliparous or late parity
Fibrocystic disease of the breast
Female relatives with carcinoma of the breast
Increasing age

Although numerous medications including tranquilizers,
steroids, thyroid preparations, and antihypertensives have been
suggested, little data support any causal relationship. Obesity has
been implicated, probably because of fat's ability to increase the
conversion of estrogens into active metabolites.

The tumor is typically discovered as a breast lump on examination by the patient or physician. Mammography is a complemen-
tary technique revealing additional lesions and is especially helpful
when the breasts are difficult to examine, i.e., multicystic or
pendulous. (Further discussion of mammography and screening
recommendations for women over the age of 50 can be found in
Chapter 14.) Mammographic signs of malignancy include stellate
masses and microcalcifications. Pathological examinations of
masses or suspicious mammographic shadows should be performed
on all elderly women without delay because of the high incidence
of malignancy at this time of life. Aspiration needle cytology is a
highly specific and sensitive test in experienced hands. Analysis of
fluid may help distinguish malignant from benign lesions. Because
of the low false-negative rate (<5%), suspicious masses should still
be subjected to intraoperative excisional biopsy. (See table on p.
257.)

In most cases, the optimal surgical therapy is that which

Malignant and Benign Lesions of the Breast*

Benign	Comments	Malignant	Frequency	Comments
Cysts	Aspiration for large, uni-locular cysts	Paget's disease	3%	Nipple itching; late metastases
Fat necrosis	Tenderness over injury site; biopsy typically done	Ductal carcinoma	70%	Common; poor prognosis if infiltrating
Fibroadenoma	Does not require excision	Medullary carcinoma	6%	Improved prognosis
		Lobular carcinoma	4%	Good prognosis; may be *in situ*
Intraductal papilloma	Frequent cause of bloody nipple discharge	Papillary carcinoma	1%	Rare
		Inflammatory carcinoma (*peau d'orange*)	1%	Not curative; palliative therapy only

*Signs of malignancy: skin dimpling, nipple retraction, fixation of lump to chest wall or skin, unilateral arm lymphedema, matted axillary nodes.

removes all clinical disease and evaluates the axilla for the presence
of tumor. Choices include lumpectomy with axillary dissection or
modified radical mastectomy for stage I patients; modified radical
or extended radical mastectomy for stage II patients with central or
medial disease (with higher rates of positive internal mammary
nodes); and modified radical or radical mastectomy for patients with
large stage II tumors or clinically positive axilla. Tissue that is not
formalin immersed should be obtained for determination of estrogen
and progesterone receptors. In a number of large surgical series,
elderly women do not show increased mortality following
mastectomy.

Staging of Breast Cancer

Tumor

1 Mass <2 cm
2 2 cm $\leq$ mass <5 cm
3 Mass $\geq$ 5 cm
4 Fixation to chest wall or skin extension; inflammatory
  carcinoma

Node

0 No nodes
1 + Ipsilateral
2 Matted ipsilateral nodes
3 Supraclavicular nodes or ipsilateral arm edema

Metastasis

0 None
1 Present

Stage

I T_1,N_0,M_0
II $T_2,N_0/N_1,M_0$
   T_1,N_1,M_0
III $T_3,N_0/N_1,M_0$
IV Any T,N_3,M_0
   Any T,Any N,M_0
   T_4, Any N,Any M_1

The 5-year survival for malignant lesions without axillary lymph node involvement is 80% for stage I; 55% for stage II; and 35% for stage III. Rapid rehabilitation of the involved shoulder girdle, psychosocial support, and biannual examination of the contralateral breast for recurrence with annual mammography are indicated.

Elderly patients with axillary lymph node involvement have significant relapse rates and should be considered candidates for postoperative adjuvant treatment. The best predictor is the number of involved nodes; 5-year survival is 60% for one to three positive axillary nodes, falling to 30% with four or more nodes positive. Treatment options include radiation therapy, hormonal manipulation, or chemotherapy. Radiotherapy improves local control of disease but does not enhance survival. Radiation is equally effective at alleviating symptoms if initiated when local disease becomes evident. Typical indications include painful bony metastasis, chest wall skin nodules, arm lymphedema from regrowth of involved nodes, or increased intracranial pressure from cranial metastasis.

Over half of postmenopausal women with breast cancer display estrogen and/or progesterone receptor positivity. Initial prognosis is improved both for overall survival and recurrence-free interval in this group. Estrogen-receptor-positive tumors will respond in 60% of cases to hormonal manipulations; metastasis to soft tissue, bone, or lung is most likely to improve. The drug of choice is tamoxifen (10–20 mg b.i.d. p.o.), an antiestrogen.

Toxicity is minimal and may include symptoms of nausea, transient cytopenias, and an occasional "flare" early in the course of therapy when given for bone pain. Survival rates, quality of life, and disease-free interval are improved. Addition of an antiandrogen (fluoxymesterone) or medical adrenalectomy agents (aminoglutethimide) have been effective in relapsing cases. At the present time, luteinizing hormone-releasing hormone (LHRH), although anecdotally beneficial, remains experimental.

Postmenopausal women with estrogen-receptor-negative tumors have a minimal response rate (10%) to hormonal manipulations. Chemotherapy should be given to these patients when positive lymph nodes are found. Cyclophosphamide, methotrexate, and 5-fluorouracil combination (CMF) regimens are the most commonly used. Myelogenous suppression and nausea are often the limiting toxicities at all ages. Survival and 5-year relapse-free rates in postmenopausal patients by dose response are listed. Nodal status is still prognostically significant (not shown).

CMF Adjuvant Chemotherapy for Breast Cancer in Post Menopausal Women

Chemotherapy status	Five-year freedom from relapse	Five-year survival
Control	49%	71%
≥85% dose received	75%	77%
<85% dose received	52%	62%

Every effort should be made to administer the entire chemo-therapy dosage prescribed, even if a time delay is required. For recurrent or resistant disease, doxorubicin currently appears to be the best single agent, providing a 6- to 9-month response in as many as 30% of cases.

VII. CARCINOMA OF THE LUNG

Lung cancer peaks in incidence during the seventh decade of life. In 1985, death rates from lung cancer were 55/100,000 in men and 18/100,000 in women. Only the rare alveolar cell carcinoma lacks a smoking connection; other implicated environmental toxins include asbestos, arsenicals, and uranium. The steadily increasing incidence with age appears to reflect the cumulative lifetime ex-posure to smoking and other environmental agents. Unfortunately, because of the nonspecificity of symptoms, most patients (70%) present only after spread has occurred. Screening programs and preventative health measures to decrease smoking are essential.

Signs and Symptoms of Lung Cancer

Extension of Tumor Mass

Pleuritic pain or effusion (lymphangitic obstruction)
Pain (rib invasion)
Tracheal deviation
Superior vena cava syndrome (facial plethora)
Horner's syndrome (sympathetic chain involvement)
Phrenic nerve paralysis with elevated hemidiaphragm

(continued)

Signs and Symptoms of Lung Cancer (*Continued*)

Extension of Tumor Mass (*Continued*)

Hoarseness (recurrent laryngeal nerve palsy)
Pericardial extension
Pancoast's syndrome (eighth cervical/first thoracic nerve
  palsies)
Dyspnea (diffuse metastasis or lymphangitic spread)

Primary Tumor Mass

Cough
Hemoptysis
Wheezing
Dyspnea
Pneumonitis
Infection

Indirect Effects

SIADH
ACTH-like production
Eaton–Lambert syndrome
Hypertrophic osteoarthropathy

Evaluation of a lung mass requires tissue examination. Percutaneous fine-needle aspiration for cytological studies has a specificity of 85–90% with only a 2–4% false-positive rate and is an excellent choice for peripheral lesions. Flexible fiberoptic bronchoscopy with brushings and washings of the involved lobe has a specificity of 80% but a higher false-positive rate. Direct biopsy may be possible for proximal endobronchial or invasive lesions. Thoracotomy is the gold standard procedure, revealing a 50–60% malignancy rate for solitary pulmonary nodules in patients over 50 years of age.

Prognosis is strongly influenced by stage and histological type. Small-cell carcinoma is rarely resectable at the time of diagnosis and typically has metastasized. Chemotherapy is the preferred treatment for this tumor type. Typical sites of metastases include liver, brain, marrow, and bone. Non-small-cell types show the best prognosis for the slow-growing and late-metastasizing squamous-cell type, but resectability is often difficult because of the proximal location of these lesions. Treatment optimally includes lobectomy or pneumonectomy if required; "curative doses" of radiation have

Pathology and Prognosis of Lung Cancer by Type

Type	Percent proximal	Percent of total	Percent resectable	Five-year survival if resected
Squamous cell	80%	40–60%	60%	37%
Adenocarcinoma	30%	20–25%	38%	27%
Large cell (or un-differentiated adenocarcinoma)	50%	10–15%	38%	27%
Small cell	85%	20–25%	10%	0%

had poor results compared with surgery. Noncurative radiation has been used for palliation of obstructive or pain symptoms and as adjunctive therapy in resected patients having positive mediastinal nodes. This may increase survival in selected patients, particularly those with a squamous-cell morphology.

Lung Cancer: Staging

Tumor

0 Cytology, no pulmonary nodule
1 Mass ≤3 cm
2 Mass >3 cm, not closer than 2 cm to the carina; any size mass extending to visceral pleura
3 Any size mass within 2 cm of the carina or extending to parietal pleura, diaphragm, or mediastinum

Nodes

0 None
1 Ipsilateral hilar nodes involved
2 Mediastinal involvement

Metastasis

0 None
1 Present

Stages

I $T_1, N_0/N_1, M_0$
T_2, N_0, M_0
II T_2, N_1, M_0
III T_3, Any N, Any M
Any T, N_2, Any M
Any T, Any N, M_1

Prior to any pulmonary resection, a preoperative pulmonary evaluation is extremely important, particularly in elderly smokers. Preferably, tests should be done following abstinence from smoking and treatment of any underlying infections. Preoperative hypercapnia, an $FEV_1 \leq 1$ liter, or $FEV_1 > 50\%$ of vital capacity typically makes surgery less desirable. Selective xenon ventilation studies of the noninvolved lung may reveal how much ventilation will remain postoperatively. This may be of major importance in marginal cases.

Chemotherapy for small-cell carcinoma has been effective in prolonging the dismal survival rates (50% survival less than 6 months) and in some cases providing long-term remission. Functional status appears to predict survival more accurately than absolute age. High-dose, multiagent chemotherapy protocols are used, typically combining cyclophosphamide, adriamycin, vincristine, VP-16, methotrexate, or daunorubicin. Limited radiation to stage I lesions may also increase response rates. Following induction, the patient should be restaged and receive prophylactic radiation therapy to the brain with the maintenance regimen.

VIII. GASTROINTESTINAL MALIGNANCY

A. Cancer of the Colon and Rectum

This form of cancer is surpassed only by lung cancer as a cause of cancer deaths. Incidence peaks between the ages of 75 and 85 with rates of 100–120/100,000. Similar to diverticulosis, the incidence of colorectal malignancy is strikingly higher in Western society and has been largely attributed to differences in diet. Low fiber and high animal fat in Western diets have been epidemiologically implicated.

Colorectal Carcinoma: Common Signs and Symptoms

1. Change in bowel habits: frequent constipation, diarrhea, or correction of preexisting constipation
2. Change in stool diameter
3. Rectal bleeding
4. Anemia
5. Vague abdominal pain
6. Abdominal mass
7. Intestinal obstruction

Most tumors are adenocarcinomas originating in adenomatous or villous polyps. The distribution of polyps and therefore carcinomas is outlined. The arbitrary breakdown at 25 cm and 60 cm from the anus reflects the maximum reach of the rigid proctosigmoidoscope and the flexible sigmoidoscope, respectively. All polyps over 1 cm should be removed.

Duke's Classification Stages for Colorectal Cancer

Stage	Description	Post-operative 5-year survival
A	Limited to mucosa	88%
B	Limited to bowel wall	60%
C	Local lymph nodes involved; invasion outside bowel wall	20%
D	Distant metastasis	0%

The digital examination with testing for occult blood should be done routinely. Proper preparation of the bowel for sigmoidoscopy or barium studies is imperative, but caution is advised to prevent electrolyte and fluid losses when using strong cathartics in the elderly. In general, a single Fleet's enema will be sufficient. Following barium studies, a mild laxative should be given immediately to prevent impaction.

Prognosis depends on depth of invasion and, to a lesser degree, on histological differentiation. (See table on p. 265.)

Although postoperative radiation therapy in rectal carcinoma appears to delay the recurrence of local disease, it has no effect on survival rate. Complications including radiation enteritis must be weighed against potential benefits. The role of preoperative radiation therapy and combined chemotherapy adjunctive protocols in rectal carcinoma are being evaluated. The only agent with demonstrated effectiveness in colon carcinoma is 5-fluorouracil (5-FU); approximately 20% of patients with metastatic disease respond. Efficacy of infusing 5-FU into the hepatic vasculature is currently being evaluated in several medical centers for those with existing liver metastases.

B. Carcinoma of the Stomach

Fortunately, this type of cancer has been decreasing in incidence in the United States. Despite this, it still remains a major

Colorectal Polyps and Adenocarcinoma: Distance from Anus

Polyp	Incidence	Invasive malignancy	Site 0–25 cm All benign polyps	Site 25–60 cm All benign polyps	Site 60 cm All benign polyps
Adenomatous	75%	5%	5%	61%	34%
Mixed elements	15%	22%	5%	61%	34%
Villous	10%	40%	27%	36%	36%
Frank carcinoma	—	(100%)			

problem, with incidence rates in men of 150/100,000 at age 70. Presenting symptoms are usually visually vague and ill-defined in the elderly, leading to delays in diagnosis with a resultant high mortality. Increased physician and patient awareness of the disease in Japan, along with early use of contrast barium studies and endoscopy, has greatly increased early diagnosis in that country. There is a significant association with atrophic gastritis and pernicious anemia.

Features of Gastric Carcinoma in the Elderly

1. More common in men (2 : 1, men–women ratio)
2. Most common between 70 and 75 years of age
3. More common in patients with blood group A
4. Associated with atrophic gastritis/pernicious anemia

Gastric Carcinoma: Signs and Symptoms

Anorexia
Anemia
Nausea, vomiting
Dysphagia
Postprandial fullness or bloating
Weight loss/cachexia
Epigastric pain: "dull or boring"; may be partially relieved by antacids
Blood loss: acute or chronic

Surgical techniques and required resection are determined by the tumor location in the stomach, paths of locally and regionally draining nodal networks, extent of extension into esophageal lymphatics in proximal lesions, and local invasion of other organs such as pancreas and spleen. The tumor and draining nodes should be removed *en bloc* if possible. Patients who are properly resected by a distal partial gastrectomy have less difficulty with nutrition and weight maintenance postoperatively than do patients who undergo proximal subtotal or total gastrectomy. Chemotherapy, including 5-fluorouracil and methyl-CCNU or mitomycin, has been found to improve survival in patients with advanced disease and may additionally be used as adjuvant therapy in patients with completely resected tumor mass.

C. Pancreatic Carcinoma

Malignancy of the exocrine pancreas is rarely diagnosed in curable stages. Typical patients present with advanced local disease, abdominal pain, and weight loss. Biopsy should be performed at operative evaluation to rule out less aggressive islet cell tumors, which can be resected and treated with chemotherapy (streptozotocin). Resection with biliary and bowel bypass should only be performed for pancreatic adenocarcinoma less than 2 cm in size. The best prognosis is found in those with small tumors isolated to the head of the gland. In these cases, presenting signs usually appear early because of painless jaundice resulting from ductal compression. Median survival in these cases is over 1 year; good functional status, regardless of age, is a favorable prognostic indicator. Diagnosis and staging are best made with endoscopic retrograde cholangiopancreatography and abdominal CT scan.

D. Carcinoma of the Esophagus

The incidence of carcinoma of the esophagus rises steadily with increasing age and must be considered in any elderly person complaining of dysphagia. Pain and discomfort are a late feature. Diagnosis must always be confirmed by endoscopic biopsy. Bronchopulmonary symptoms commonly result in untreated patients as a result of fistula formation between the esophagus and trachea. Surgery should be attempted if the site of lesion allows. Although radiation therapy may be used in inoperable patients, it is rarely curative. Pain control and nutritional support are essential.

Features of Carcinoma of Esophagus in the Elderly

1. Most frequent between the ages of 65 and 75; peak incidence 40/100,000 at age 75
2. More common in men
3. Patients with Plummer–Vinson syndrome (iron deficiency anemia with esophageal web), alcoholism, and smoking have highest incidence
4. Dysphagia for solids precedes dysphagia for liquids
5. Treatment is by surgery or radiotherapy
6. Prognosis is poor; 5-year survival rate is less than 5%

IX. THYROID CANCER

This subject is discussed in Chapter 7.

X. ADENOCARCINOMA OF THE PROSTATE

This cancer is extraordinarily prevalent in adult men. On the basis of autopsy studies and pathological examination of tissue removed for presumed benign prostatic hypertrophy, a prevalence of 20–25% at age 64, rising to almost 50% of all men by age 80, has been noted. As many as 40% of cases are clinically unsuspected prior to surgery. Prostate examinations must be routinely done. Even then, however, stage I lesions are often difficult to recognize clinically. Unfortunately, cure is only available for patients with stage I and ll lesions, so examination for palpable prostate nodules should be performed frequently; there should be an early referral to a urologist in those cases of difficult-to-examine or questionable glands.

Evaluation of Adenocarcinoma of the Prostate

Symptoms	Pathological Diagnosis	Staging Evaluation
1. Obstructive: hesitancy urgency frequency weakened stream dysuria	1. Prostate needle biopsy (core or fine-needle aspirate)	Prostatic acid phosphatase Bone scan Chest X ray If above negative and nodule palable (stage B or C): pelvic lymphadenectomy or lymphangiography ± abdominoperineal CT with needle biopsy of enlarged nodes
2. Tumor spread: low back pain pathological fractures	2. Transurethral resection	

A. Staging

Prognosis in prostate carcinoma clearly depends on stage, tumor differentiation, and degree of hormonal dependence. Tumors confined within the prostatic capsule are theoretically curable by surgical resection; microscopic extension is frequently present, however, in stage B_2. Small, well-differentiated tumors (A_1) frequently (>75%) represent a "pathologist's diagnosis" only, with no potential for progressive growth. Poorly differentiated, aggressive tumor lines are selected by treatment and tend to be testosterone-independent clones. A very poor prognostic finding at diagnosis is an extremely low serum testosterone, implying tumor growth despite absent testosterone stimulation.

Staging of Prostate Carcinoma

Stage	Definition	Comment
A_1	Nonpalpable, ≤5% of gland involved	Typically well-differentiated; 10-year cancer death rate <5% (untreated)
A_2	Nonpalpable, >5% of gland involved	May be poorly differentiated; 10-year cancer death rate 20% (untreated)
B_1	Palpable nodule in one prostatic lobe	Some authors require size <1.5 cm; 5% concomitant invasion of seminal vesicles
B_2	Palpable nodule in both prostate lobes	Some authors include nodules >1.5 cm; 50% concomitant invasion of seminal vesicles
C	Palpable extracapsular extension	Acid phosphatase ±; 10-year cancer death rate 75% (untreated)
D_1	Metastatic disease to lymph nodes	Acid phosphatase usually +; prostate any size
D_2	Metastatic disease to other metastatic sites	Acid phosphatase elevated; prostate any size; 10-year cancer death rate <90% (untreated)

B. Therapy

Treatment of encapsulated prostate carcinoma is directed toward potential cure. Simple prostatectomy or TURP is sufficient for the majority of well-differentiated A_1 tumors. Progression of A_2 after TURP, however, suggests the need for further adjuvant treatment in the form of a more radical second surgery (technically difficult following a recent simple prostatectomy) or radiation therapy. External beam radiation is available to deliver doses of

6000–7000 rads in the prostatic area; any involved lymph nodes are not treated effectively by this method. Up to 20% of stage A_2 lesions will have microscopic metastases on lymph node dissection. Stage B carcinomas (particularly B_1) have a more favorable response to radical prostatectomy than to external beam irradiation, again because of improved metastatic control. Local (intraprostate) disease control is equivalent. The toxicity profile of the two procedures is very different, however, and should be considered in those elderly patients otherwise thought to have shorter life expectancies. Node dissection to confirm true stage B status followed by radioactive iridium implants has been evaluated and can give excellent local control with less impotence and incontinence.

Comparative Toxicities of Pelvic Radiotherapy and Radical Prostatectomy

Toxicity	Radiotherapy	Radical prostatectomy
Incontinence	5–10%	5–50%
Impotence	40%	90–95%
Structure	5%	3–20%
Acute enteritis	30–50%	—
Chronic GI adhesions	10%	—
Rectal injury	—	0.5–5%
Thromboembolism	—	10%
Lymphedema	10–30%	—

Although stage C disease is not curable, radiation therapy may be used for palliation of local pain, obstruction, or distant metastasis to bone or spinal cord. The effectiveness of hormonal manipulation is disputed; clearly, no prolongation of life is achieved. Because of the typical emergence of hormonal independence following androgen suppression for 0.5–1.5 years, this treatment should be withheld until symptoms require palliation.

Stage D disease should be treated conservatively. A TURP is indicated only for obstruction; radiotherapy should be reserved only for serious metastasis, and hormonal manipulations used only when thought to be of use in limiting symptoms. Standard therapy has either included bilateral orchiectomy or diethylstilbestrol (DES, 1 mg T.I.D.); in both cases, the effect of testicular testosterone is blocked, but adrenal androgen production is preserved. Newer trials suggest the use of luteinizing hormone-releasing hormone (LHRH) agonists to block LH secretion and decrease testicular testosterone, eliminating the cardiovascular and feminizing effects of DES or the surgical morbidity of castration. Side effects are limited to androgen deficiency (hot flashes, erectile impotence) and to early LH release causing bone pain flare for the first week of therapy. Ninety percent of patients respond; duration of response is similar to DES. The LHRH is delivered subcutaneously (Leuprolide® s.c. q.d.) or intramuscularly (Zoladex® i.m. q28d). Combination with an anti-androgen (Flutamide®) may block the effects of residual adrenal androgens and possibly flare effects; evaluation studies regarding combination therapy are in progress.

XI. GENITOURINARY MALIGNANCY

Bladder carcinoma typically presents with hematuria; pathological diagnosis is made by cystoscopic biopsy. Carcinoma *in situ*, or stage I, may be treated by multiple resections, fulgurations, and intravesical chemotherapy instillation. Submucosal invasion is treated by preoperative radiation and radical cystectomy with ureteral diversion into an ileal conduit. Residual tumor following radiation constitutes the subgroup benefiting from postoperative chemotherapy with cisplatin, methotrexate, or daunorubicin. An association with smoking seems likely; data suggest that vitamin A may support normal differentiation.

Renal cell adenocarcinoma is three times more frequent in men. Although incidence peaks during the fifth decade of life, it is found commonly in older age groups. Epidemiologic studies have implicated both cigarette smoking and thorotrast exposure, a contrast medium used in the 1920s. Nonspecific symptoms lead to delayed diagnosis; metastatic disease is present in over 50% of patients at the time of presentation.

Regardless of the primary presenting symptom, patients have a 30–40% 5-year survival rate. Presentation caused by metastatic manifestations is associated with a <5% 5-year survival; presymptomatic patients discovered during other evaluations have a >60% 5-year survival. Clinical stage further stratifies survival rates. Evaluation of a renal space-occupying lesion should include sonography to evaluate cystic characteristics and an intravenous pyelogram to clarify vascularity. Although most renal cell carcinomas are hypervascular and solid (85%), greater numbers of early, asymptomatic tumors are hypovascular (10–20%). Cyst punc-

Symptoms of Renal Adenocarcinoma

Symptoms/signs	Percent at time of nephrectomy	Percent surviving 5 years
Hematuria	59	40
Acute varicocele (L > R)	2	43
Pain (abdominal)	41	44
Anemia of chronic disease	21	38
Diffuse renal calcification	13	46
Abdominal complaints or mass	45	35
Fever	7	38
Erythrocytosis	21	40
Classic triad (flank pain, gross hematuria, abdominal mass)	9	31
Hypercalcemia	3	35
Metastasis (lung, lymph node, liver, bone, brain)	10	3

ture can confirm a benign cyst, with fluid containing a low LDH, protein, fat, and amylase content; angiography should be performed preoperatively to exclude renal vascular anomalies.

Radical nephrectomy (removing the kidney *en bloc* with adjacent fascia) has improved survival, particularly for stages II and III.

Adjuvant postoperative radiation has been attempted, particularly in stage II and IIIA patients, with variable response. No clearly effective chemotherapeutic regimens have been established

Staging of Renal Adenocarcinoma

Stage	Description	Nephrectomy: 5-year survival	Radical nephrectomy: 5-year survival
I	Intrarenal	77%	87%
II	Renal, capsule and adjacent Gerota's fascia	31%	64%
IIIA	Invasion of renal vein ± vena cava	8%	40%
IIIB	Invasion of regional lymph nodes		
IIIC	IIIA and B		
IV	Organ invasion or distant metastasis	5%	5%

to date; recent data suggest, however, that megestrol acetate (Megace®) may limit tumor growth in metastatic disease.

XII. GYNECOLOGICAL MALIGNANCY

Uterine adenocarcinoma is the most common gynecological malignancy of elderly women; despite this, it accounts for less than half of all gynecological tumor deaths (see ovarian carcinoma, below). As in breast cancer, the major risk factor appears to be unopposed estrogen effects. It is essential that perimenopausal estrogen replacement be given cyclically or include cyclic progestins. Endometrial gland hypertrophy in response to estrogen presence leads to peri- or postmenopausal bleeding; this should be distinguished from malignancy resulting from atrophic bleeding. Gynecological evaluation must be a routine part of medical care. Uterine cytology via suction or dilatation and curettage may be necessary. If a lesion is noted, hysterectomy and oophorectomy are highly successful for limited stages. Radiation therapy may be added as adjunctive treatment in invasive stages; progestational agents and single-agent chemotherapy (doxorubicin) have been successful in palliating metastatic symptoms.

Ovarian carcinoma is typically discovered late in the course of illness because of nonspecific early symptoms. Very rarely, an enlarged ovary on bimanual examination will lead to ultrasonographic investigation and diagnostic laparoscopy. Typically, the patient presents with abdominal mass or ascites. Disease limited to one or both ovaries is treated by total hysterectomy and bilateral salpingoophorectomy; later stages with pelvic and abdominal extension receive chemotherapy with melphalan combinations.

XIII. MISCELLANEOUS TUMORS

Brain tumors have two incidence peaks: one in early infancy/childhood and another in late adulthood. Adult brain tumors occur mainly above the tentorium, with the most frequent being a relatively low-grade, noninvasive astrocytoma. Higher-grade tumors with proliferation of glial elements, known as glioblastima multiforme, peak in incidence at age 70. Symptoms at presentation are generalized (seizure, headache) or reflect the local function of the affected brain (weakness, aphasia). There are a slight male preponderance and higher rates in whites (3–5/100,000 at age 70). Rate of cure by resection is good in cases of encapsulated tumor involving nonessential, resectable brain areas. High-grade tumors respond palliatively to irradiation.

Head and neck tumors are primarily squamous cell carcinomas, which spread locally to nodes and less commonly cause distant metastasis. Implicated etiologic factors include smoking, alcohol, or snuff/chewing tobacco usage. Premalignant dysplastic areas of leukoplakia (whitish) and erythroplasia (velvety red) should be screened for and, if present, treated aggressively. Chronically inflamed areas from ill-fitting dentures should also be biopsied and, if present, treated. Surgery, radiation, or a combination of both is highly effective for localized lesions.

Melanoma and skin cancer are discussed in Chapter 8.

SUGGESTED READING

Austin-Seymour M. M., Hope R. T., Cok R. S., et al: Hodgkin's disease in patients over sixty years old. *Ann Intern Med* 100:13–18, 1984.

Baker L. H.: Breast cancer detection demonstration project: Five year summary report. *CA* 34:194–225, 1982.

Bolin R. W., Robinson W. A., Sutherland T., et al: Busulfan versus hydroxyurea in long term therapy of chronic myelogenous leukemia. *Cancer* 50:1683–1686, 1982.

Ciampi A., Bush R. S., Gospodarowicz M., et al: An approach to classifying prognostic factors related to survival experience for nonHodgkin's lymphoma patients based on a series of 982 patients: 1967–1975. *Cancer* 47:621–627, 1981.

Cohen H.J., Bartolucci A.: Age and the treatment of multiple myeloma; The Southeastern Cancer Study Group experience. *Am J Med* 79:316–324, 1985.

DeWys W. D., Killen J. Y.: The paraneoplastic syndromes, in Rubin P. (ed): *Clinical Oncology for Medical Students and Physicians: A Multidisciplinary Approach* Washington, American Cancer Society, pp. 112–119, 1983.

Gale R. P., Foon K. A.: Chronic lymphocytic leukemia: Recent advances in biology and treatment. *Ann Intern Med* 103:101–120, 1985.

Goldstein S., Reis R.: Mechanisms of cellular aging. *Ann Intern Med* 102:218–228, 1985.

Kahn S. B., Begg C. B., et al: Daunorubicin, Ara-C, and 6-thioguanine in acute myeloblastic leukemia in the elderly. *J Clin Oncol* 2:865–870, 1984.

Kalser M. H., Barkin J., MacIntyre J. M.: Pancreatic cancer: Assessment of prognosis by critical presentation. *Cancer* 56:397–402, 1985.

Mayer R. J.: Acute lymphoblastic leukemia in adults. *Ann Intern Med* 101(4):552–554, 1984.

McLendon R. E., Robinson J. S., Chambers D. B., et al: The glioblastoma multiforma in Georgia 1977–1981. *Cancer* 56:894–897, 1985.

Minna J. D., Higgins G. A., Glatstein E. J.: Cancer of the lung, in DeVita V.,

Hellman S., Rosenbery S.(eds): *Cancer: Principles of Oncology*. Philadelphia, Lippincott, 1982, pp 396–474.

Paulson D. F., Lin G. H., Hinshaw W., et al: Radical surgery versus radiotherapy for adenocarcinoma of the prostate. *J Urol* 128:502–504, 1982.

Paulson D. F., Perez C. A., Anderson T.: Genitourinary malignancies, in DeVita V., Hellman S., Rosenbery S. (eds): *Cancer: Principles of Oncology*. Philadelphia, Lippincott, 1982, pp 732–750.

Reiffers J., Raynal F., Broustet H.,: Acute myeloblastic leukemia in elderly patients: Treatment and prognostic factors. *Cancer* 45:2816–2820, 1980.

Westbrook C. A., Groupman J. E., Golde D. W.: Hairy cell leukemia: Disease pattern and prognosis. *Cancer* 54:500–506, 1984.

11

Nutrition and the Geriatric Patient

I. CAUSES OF MALNUTRITION IN THE ELDERLY

Nutritional problems are common in both healthy, community-dwelling elderly and hospitalized and institutionalized geriatric patients.

Factors Involved in the Development of Malnutrition in the Elderly

1. Physical impairments
   Poor vision
   Poor dentition/dentures
   Arthritis
   Immobility
2. Physiological impairments
   Malabsorption and maldigestion
   Loss of taste and smell
3. Pathological conditions
   Dementia
   Depression
   Disease states: cancer, parkinsonism, hypothyroidism,
      atherosclerosis
4. Social factors
   Poverty
   Alcoholism
   Poor dietary habits
   Isolation
5. Iatrogenic causes
   Drug–nutrient interactions
   Prescribed diets

Iatrogenic causes of nutritional insufficiency, including dietary restrictions and drug–nutrient interactions, may frequently be overlooked.

Diets with Potential for Adverse Effects

Diet	Potential Problems
Lactose-free diet	Suboptimal calcium intake; subsequent osteoporosis
Low-protein diet	May lead to protein malnutrition with hypoalbuminemia and muscle wasting
Low-salt diet	Food apathy and decreased nutrient intake secondary to lack of taste
High-fiber diet	Bloating; gas and/or abdominal distension

A number of factors predispose the elderly to drug–nutrient interactions; these include the following:

1. The elderly are the largest users of drugs in our society.
2. Multiple drug use is higher in the elderly.
3. Prolonged or chronic drug therapy is more common in the elderly.
4. Drug absorption, distribution, and metabolism may be altered in the elderly.
5. Marginal nutrient intake is common in the elderly.

Potential Drug–Nutrient Interactions

Drug	Nutrient Deficiencies
Mineral oil	Vitamins A, D, K, E
Phenolphthalein	Potassium, vitamin D, calcium
Cholestyramine	Vitamin B_{12}, iron, fat, vitamins A, K, D
Colchicine	Fat, lactose, vitamin B_{12}, potassium
Phenformin	Vitamin B_{12}

(continued)

Potential Drug-Nutrient Interactions (*Continued*)

Drug	Nutrient Deficiencies
Glucocorticoids	Calcium, potassium
Antacids	Phosphate
Potassium chloride	Vitamin B_{12}
Aspirin	Iron
Sulfasalazine	Folate
Diphenylhydantoin	Vitamin D, calcium
Furosemide	Calcium, potassium, magnesium, zinc, water, sodium
Penicillamine	Zinc, copper, vitamin B_6
Isoniazid	Vitamin B_6
Warfarin	Vitamin K
Digoxin	Protein calories

II. RECOGNITION OF MALNUTRITION IN THE ELDERLY

Recognition of nutritional insufficiency is more complicated in the elderly because of numerous age-related changes in body composition and physiology. In fact, both the aging process *per se* and malnutrition may have parallel effects on a variety of physiological parameters.

Effect of Age and Malnutrition on Organ Function

Parameter	Aging	Malnutrition
Cardiac output	↓	↓
Basal metabolic rate	↓	↓
Renal blood flow	↓	↓
Glomerular filtration rate	↓	↓
Glucose clearance	↓	↓
Cell-mediated immunity	↓	↓
Tendency for infection	↑	↑
Pulmonary functions	↓	↓
Skeletal muscle mass and strength	↓	↓
Intestinal mucosal mass	↓	↓

Disease processes that occur with increased frequency with age may also impair the ability to assess nutritional status. For example, edema caused by heart failure could confuse interpretation of body weight. Likewise, cutaneous anergy is seen both with increasing malnutrition and with some cancers, infectious processes, and aging itself. Age-related changes in the distribution of body fat and skin turgor may make anthropometric measurements such as a triceps skinfold thickness an unreliable index of total body fat. The creatinine–height index, a standardized measurement of protein nutrition as reflected by skeletal muscle mass, must be adjusted for age-associated reductions in both musculature and creatinine clearance.

Although these and other factors must be considered in the nutritional assessment of geriatric patients, several general principles still apply.

III. PRINCIPLES OF NUTRIENT ASSESSMENT

In the absence of fluid changes, progressive loss of body weight signifies caloric insufficiency. The involuntary loss of greater than 10% of body weight should be considered highly significant.

Although an imperfect parameter, serum albumin tends to correlate with visceral protein status in the elderly. The serum albumin may remain normal, however, in some cases of protein–calorie malnutrition (marasmic type). Therefore, a normal albumin does not necessarily mean that a patient is not suffering from malnutrition. Likewise, factors other than nutrition *per se* may affect albumin levels, including liver disease, state of hydration, and abnormal losses, as occur in the nephrotic syndrome. Nevertheless, a significantly diminished serum albumin (<3.5 g/dl) should not be considered a "normal" part of aging; protein malnutrition must be considered.

When applying formal nutrition assessment parameters to the elderly, including height, weight, triceps skinfold, midarm muscle circumference, and creatinine–height index, only age-adjusted standards should be used. Comparing these parameters to standards derived from younger adults can lead to significant miscalculations of nutritional status.

An abnormal hemoglobin or hematocrit in the healthy, older individual may result from nutritional insufficiencies including folate, iron, and, rarely, vitamin B_{12}. Anemia should never be considered a part of normal aging.

Frank clinical signs of vitamin or mineral deficiencies are rare in the elderly. Accurate assessment of these nutrients by laboratory testing may be indicated in some patients.

Dietary histories, particularly using short-term recall, may be misleading in the elderly. To assess food intake accurately in the institutionalized patient, formal "calorie counts" performed by a registered dietician are often invaluable.

IV. MANAGEMENT OF THE MALNOURISHED

Potentially Reversible Factors Contributing to Malnutrition

Problem	Comment
Oral environment	Therapeutic oral and dental procedures should be considered; approximately 50% of Americans over age 65 are edentulous
Psychological factors	Management of depression, when present, is essential; eating is a social function that may be affected by isolation, loneliness, and depression
Systemic complaints	Control of pain, nausea, constipation, or other symptoms that could affect appetite must be promptly treated; timing the use of analgesics or antinausea medicines to coincide with meals, rather than giving them PRN, is sometimes beneficial
Disease states	Evaluate for underlying disease states that increase in frequency with aging

(continued)

Potentially Reversible Factors Contributing to Malnutrition (*Continued*)

Problem	Comment
	and could affect food intake, digestion, or absorption; some examples include hypothyroidism, parkinsonism, underlying malignancies, and atherosclerosis, including ischemia of the bowel
Social factors	Specifically ask about and intervene in adverse social conditions limiting food intake such as poverty (Is the patient eligible for food stamps?)

Carefully review the present diet with appropriate counseling to increase the spontaneous intake of food. This may include:

1. Physical consistency of foods. Does the patient have trouble chewing or swallowing certain items?
2. Caloric density and nutrient value of foods. Is the patient ingesting foods and liquids of little or no nutritional value?
3. Taste and ethnic preferences of the patient. Hypogeusia (diminished taste) is common in the elderly. Taste buds decline with age from approximately 275 to fewer than 100.
4. Prescribed diets. Certain disease states mandate limited food choices. Review their necessity and discuss alternatives.
5. Use of dietary supplements. Make sure the patient is not ingesting large amounts of potentially harmful vitamin or mineral supplements such as vitamin A or selenium.

When giving advice regarding diet, it is best to revise the individual's customary diet rather than to try to remake a diet in an "ideal" manner. In general, assuring a varied intake from the different food groups should help reduce the likelihood of specific deficiencies.

Four Main Food Groups

1. Dairy products: milk, cheese, ice cream, and yogurt
2. Meats and other high-protein foods, including poultry, fish, eggs, and legumes
3. Fruits and vegetables, including fruit juices
4. Breads and cereals, grains, and rice

Elderly people should eat two servings daily of dairy products, one to two servings of the meat group, four servings of fruits and vegetables, and four servings of cereals. In general, the elderly should not allow poor sources of essential nutrients, such as alcohol and purified fats and carbohydrates, to replace foods from the basic food groups.

V. FIBER

Fiber usually refers to the nondigestible polysaccharide components of foods and one noncarbohydrate, lignin. Fiber holds water, resulting in bulkier and often softer bowel movements. A diet high in fiber may reduce constipation as well as alleviate some of the symptoms of the irritable bowel syndrome and diverticulosis. In addition, certain fibers may help reduce serum cholesterol, improve glucose tolerance, and even play a role in the development or prevention of colon cancer. Although there is no definitive recommendation for fiber intake requirements, an intake of approximately 30 g per day is probably safe and adequate for most healthy individuals. This can be achieved either by diet or with supplements.

Approximate Fiber Content of Selected Foods*

Grains and cereals		Fruits	
White bread	2.7	Prunes	7.7
Whole wheat bread	8.5	Banana	3.4
Pancake, waffle	0.9	Raisins	6.8
All-Bran™	26.7	Apple (peel and	
Cornflakes	11.0	flesh)	1.5
Rice Krispies™	4.5	Cherries	1.2
Special K™	5.5	Dried apricots	24.0
Puffed wheat	15.4	Orange	2.0

(continued)

Approximate Fiber Content of Selected Foods* (*Continued*)

Grains and cereals		Fruits	
White rice	0.8	Others	
Oatmeal	7.0	Peanuts	8.1
Vegetables		Peanut butter	7.6
Peas	12.0	French fries	3.2
Spinach	6.3	Lentil soup	2.2
Green beans	3.2	Strawberry jam	1.1
Corn on cob	4.7		
Cauliflower	1.8	Bulk laxatives	
Broccoli	4.1	Metamucil™	
Baked potato	2.5	(g/tsp)	3.5
Baked beans	7.3	Perdiem Plain™	
Lettuce	1.5	(g/tsp)	5.2
Cucumber	0.4	Fiber Med™	
Onions	2.1	(g/cookie)	5.0
Carrots	2.9		
Celery	1.8		

*Grams fiber/100 g edible portion.

VI. SPECIFIC NUTRITIONAL REQUIREMENTS IN THE ELDERLY

Little is known regarding actual requirements for the elderly of many nutrients, including vitamins, minerals, and trace elements. Controversy even exists over the optimal intake of macronutrients such as protein. This lack of information is reflected by the recommended dietary allowances (RDA) that have been established by the Food and Nutrition Board of the National Academy of Sciences. For most nutrients, all geriatric patients are included in one group, subdivided as either 51 years or older or as "adult." Since elderly persons represent a very heterogeneous population, this approach is simplistic. Normal age-related changes and age-prevalent illness result in numerous changes in nutrient requirements and/or utilization.

Recommended Daily Dietary Intake of Nutrients for Healthy Elderly*

Nutrient	Age	Male	Female
Calories (kcal)	51–75	2000–2800	1400–2000
	76+	1650–2450	1200–2000
Protein (g)	51+	56	44
Vitamin A (μg retinol equivalents)	51+	1000	800
Vitamin D (μg)	51+	5.0	5.0
Vitamin E (mg α-tocopherol)	51+	10	8.0
Ascorbic acid (mg)	51+	60	60
Thiamine (mg)	51+	1.2	1.0
Riboflavin (mg)	51+	1.4	1.2
Niacin (mg niacin equivalents)	51+	16	13
Vitamin B_6 (mg)	51+	2.2	2.0
Folacin (μg)	51+	400	400
Vitamin B12 (μg)	51+	3.0	3.0
Calcium (mg)	51+	800	800
Phosphorus (mg)	51+	800	800
Magnesium (mg)	51+	350	300
Iron (mg)	51+	10	10
Zinc (mg)	51+	15	15
Iodine (μg)	51+	150	150
Vitamin K	Adult	70–140	70–140
Biotin (μg)	Adult	100–200	100–200
Pantothenic acid (mg)	Adult	4.0–7.0	4.0–7.0
Sodium (mg)	Adult	1100–3300	1100–3300
Potassium (mg)	Adult	1875–5625	1875–5625
Chloride (mg)	Adult	1700–5100	1700–5100
Copper (mg)	Adult	2.0–3.0	2.0–3.0
Manganese (mg)	Adult	2.5–5.0	2.5–5.0
Fluoride (mg)	Adult	1.5–4.0	1.5–4.0
Chromium (mg)	Adult	0.05–0.20	0.05–0.20
Selenium (mg)	Adult	0.05–0.20	0.05–0.20
Molybdenum (mg)	Adult	0.15–0.50	0.15–0.50
Water (cc/kcal)	Adult	1.0	1.0

*Adapted from Food and Nutrition Board, National Academy of Sciences.

VII. CALORIC REQUIREMENTS

With increasing age there is a gradual fall in lean body mass, largely resulting from changes in skeletal muscle. This decrease corresponds to a decrease in basal energy requirements. Total caloric requirements, which reflect both basal expenditure and expenditure for activity, can be estimated from the basal rate. The approximate basal energy expenditure (BEE) can be calculated by using the Harris–Benedict equations:

Males: BEE = 66 + [13.7 × weight (kg)] + [5 × height (cm)] − [6.8 × age (years)]

Females: BEE = 655 + (9.6 × weight) + [1.8 × height(cm)] − (4.7 × age)

Actual caloric requirements will vary from the basal level depending on activity and concurrent illnesses. In the healthy, nonstressed, elderly patient, total caloric requirements are approximately 1.2 × BEE. Greater amounts up to 2.5 × BEE will be needed in some sick, hospitalized patients.

VIII. ENTERAL NUTRITIONAL SUPPORT OF THE GERIATRIC PATIENT

The initiation of enteral nutritional support should only follow a complete assessment to determine that the patient is receiving an inadequate oral intake. Patients with marked weight loss and/or clinical or laboratory signs of protein–calorie malnutrition almost always should be supplemented, particularly if those factors that led to the malnourished state are not readily reversible.

Medical Conditions Commonly Requiring Enteral Nutritional Support

1. Neurological/psychiatric diseases
   Cerebrovascular accidents
   Neoplasms
   Trauma
   Severe depression
   Anorexia

(continued)

Medical Conditions Commonly Requiring Enteral Nutritional Support (*Continued*)

2. Oropharyngeal/esophageal
   Obstruction
   Neoplasm
   Dysmotility
3. Gastrointestinal disorders
   Mild to moderate malabsorption
   Inflammatory bowel disease
   Some fistulas
4. Adjunct to medical/surgical treatments
   Burns
   Chemotherapy
   Radiotherapy

The enteral route should be considered first in patients with a functional gastrointestinal tract because of ease of administration, low cost, proven long-term efficacy, and good patient tolerance. Important points to consider in enteral nutrition include the route of administration, the choice of formula, and potential complications.

A. Considerations when Choosing the Mode of Administration of Enteral Supplements

Oral supplementation with a nutritionally complete formula should be utilized in patients who are motivated to increase their nutrient intake and who have an intact swallowing mechanism.

Individuals with dysphagia for solid foods as a result of esophageal strictures or tumors may be able to ingest adequate amounts of liquid supplements.

In individuals with marginal nutrient intake, substituting commercial enteral supplements or other high-calorie foods for drinks or snacks of less nutritive value (e.g., tea, diet soda, coffee) may obviate the need for further nutritional intervention.

If the spontaneous intake of adequate nutrition is not possible, tube feedings may be necessary. The most common available choices include (1) nasogastric or nasoenteric tubes, (2) gastrostomy, and (3) jejunostomy.

B. Tube Feedings

Standard surgical salem sump or other similar nasogastric tubes are rarely indicated for enteral feedings. Instead, small (6 to 10 French), soft (silicone or polyurethane) tubes specifically designed for feeding should be used. These tubes permit continued oral intake. By placing the tube distal to the pylorus, near the ligament of Trietz, the risk of pulmonary aspiration is minimal. These tubes are often weighted with mercury or tungsten to reduce accidental dislodgement.

Representative Examples of Feeding Tubes

Name	French gauge	Length (inches)	Material
Dobbhoff	8	43	Polyurethane
Keofeed	5, 7.5, 9.6	43	Silicone rubber
Entriflex	8	36 or 43	Polyurethane
Vivonex Tungsten tip	8	45	Polyurethane
Duo-Tube	5, 6, 8	40	Polymeric silicone (silastic)

In patients requiring long-term feeding, a surgically placed esophagostomy, gastrostomy, or jejunostomy tube may be worthwhile.

Advantages of gastrostomy tubes are:

1. Direct entry of foods into the stomach. The stomach's capacity as a regulator of nutrient delivery to the small bowel is maintained. This may reduce the problems of dumping syndrome or diarrhea.
2. Ease of the placement and replacement if the tube is dislodged. Gastrostomy tubes can be inserted either surgically, using local anaesthesia, or via a percutaneous endoscopic technique.
3. Large bore of tube may facilitate bolus feeding of a larger variety of foodstuffs or formula and reduces the risk of tube clogging.

Jejunostomy tubes should be considered in patients in whom the risk of or complications from aspiration are very high. Patients with absent gag reflexes, free esophageal reflux, or serious pulmonary disease may be candidates for a jejunostomy.

C. Enteral Supplements

A wide array of nutritionally complete, commercially prepared enteral supplements are available to suit almost any patient.

Products vary based on a variety of parameters, including:

1. Lactose content.
2. Osmolality.
3. Caloric density.
4. The degree of digestion or absorption required to assimilate the formula.
5. Special protein and/or electrolyte compositions for specific disease states.

Generally speaking, meal replacement formulas are polymeric mixtures of protein, fat, and carbohydrate requiring some degree of digestion for proper absorption. They are usually low residue, lactose-free, and tend to have lower osmolalities than elemental diets. Elemental formulas are monomeric solutions that often require little or no digestion for proper absorption. They are usually lactose-free, low residue, and low fat but tend to have higher osmolalities than other formulas, which could result in diarrhea, particularly if given by bolus administration. (See table on p. 290.)

D. Potential Complications of Enteral Nutrition

1. Mechanical Problems Resulting from the Tube Itself

Problem	Preventive Measures
Pulmonary aspiration	1. Check for transpyloric location of the tube, near the ligament of Trietz
	2. Elevate the head of the bed by 30°
	3. Use continuous slow-drip infusion rather than bolus administration

(continued)

Representative Enteral Nutritional Supplements

Formula	Manufacturer	kcal/ml	Protein (g/liter)	Fat (g/liter)	CHO (g/liter)	Na (meq/liter)	Osmolality (mOsm/kg)	Lactose
Blenderized tube feed								
Compleat-B	Doyle	1.07	43	43	128	56	405	+
Meal replacement								
Meritene Liquid	Doyle	0.96	57.6	32	110.4	38.2	505	+
Ensure	Ross	1.06	37.2	37.2	145	36.8	450	−
Isocal	Mead Johnson	1.06	34	44	132	23	300	−
Sustacal H.C.	Mead Johnson	1.5	61	58	190	35	650	−
Magnacal	Organon	2.0	70	80	250	43.5	590	−
Elemental								
High Nitrogen Vivonex	Norwich Eaton	1.0	45.8	0.86	208	23	810	−
Criticare HN	Mead Johnson	1.06	38	3	222	27	650	−
Vital HN	Ross	1.0	41.7	10.8	188	16.7	460	−
Special disease								
Travasorb Renal	Travenol	1.35	22.9	17.7	270.8	Negligible	590	−
Hepatic-Aid	McGraw	1.6	26.1	22.14	176.5	Negligible	900 (Mosm/liter)	−

Problem	Preventive Measures
	4. In gastric feeding, check for residual contents before administering the next feeding
Tube clogging	Periodic flushing of tube with water
Otitis media, nasal, pharyngeal, or esophageal erosions	Rare with soft, small feeding tubes and careful nursing care

2. Gastrointestinal Symptoms

Problem	Preventive Measures
Bloating, satiety, vomiting, abdominal cramps, and/or diarrhea	1. Begin feeding hypertonic solutions at a slow rate, 30 to 50 cc/hr, at 50% dilution; then increase the rate by 20 cc/hr per day until full volumes are achieved; only then increase the concentration to 75% and finally, if tolerated, to full strength
	2. Add paregoric, 5 cc q.i.d., loperamide, 2 mg q.i.d., kaolin, or other antidiarrheal agents to the regimen
	3. Reevaluate the specific choice of formula and the patient's digestive capacity; consider switching to a formula of lower osmolality, lower fat, or a lactose-free solution

3. Metabolic Complications

Metabolic complications may occur as a result of an imbalance in any of the infused nutrients contained in the enteral formula.

Because of the physiological changes that occur with aging, the elderly are particularly prone to several complications:

1. Dehydration secondary to osmotic diarrhea.
2. Fluid overload or congestive heart failure secondary to salt and water retention that accompanies carbohydrate refeeding, superimposed on the decreased cardiac index more commonly observed in the elderly.
3. Hyperglycemia secondary to infused carbohydrate is more likely in the elderly because of age-associated glucose intolerance.
4. Electrolyte disturbances, including hypokalemia, hypophosphatemia, hypomagnesemia, hypernatremia, or hypercalcemia may be noted.

IX. PARENTERAL NUTRITIONAL SUPPORT OF THE GERIATRIC PATIENT

A variety of situations frequently encountered in the hospitalized and institutionalized elderly patient may preclude adequate oral or enteral feedings.

Potential Contraindications to Oral or Enteral Feedings

1. Bowel obstruction
2. Coma
3. Some bowel fistulas
4. Intractable vomiting
5. Impaired swallowing mechanism
6. Severe malabsorption with or without diarrhea
7. Tetanus
8. Toxic megacolon
9. Jaw wired closed as in trauma cases

Depending on the severity and likely course of the underlying disease state that precludes oral intake and the nutritional status of the patient, parenteral nutritional support may be indicated. It should be considered in those situations in which maintenance or improvement in nutritional status allows for significant palliation of the underlying illness or permits proper therapy to be administered or recovery to occur. Because of the potential complications of treatment, high cost, and unproven efficacy, total parenteral nutrition (TPN) has no role in the routine postoperative care of the well-nourished geriatric patient or in an attempt to prolong the life of the terminally ill individual in whom other therapeutic options have been exhausted.

Potential Indications for TPN

1. Obstructing lesions of the GI tract where surgery must be delayed or is not possible
2. Following massive bowel resections
3. Gastrointestinal fistulas
4. Severe GI motility disturbances such as intestinal pseudoobstruction
5. Adjunctive to cancer therapy
6. Perioperative support
7. Severe hypermetabolic states following sepsis, burns, or trauma

The specific intravenous nutrient requirements in the geriatric population, particularly in those with various disease states, remain largely a matter of clinical speculation.

Suggested Daily Intravenous Nutrient Intakes for Healthy Adults

Nutrient	Intake	Nutrient	Intake
Calories	30 kcal/kg	Vitamin K	2–4 mg/week
Protein	1 g/kg	Calcium	5–20 mEq
Vitamin A	3300 IU	Phosphorus	12–36 mmol
Vitamin D	200 IU	Magnesium	8–20 mEq
Vitamin E	10 IU	Iron	1 mg
Vitamin C	100 mg	Zinc	2.5–4.0 mg
Thiamine	3 mg	Iodine	70–140 μg
Riboflavin	3.6 mg	Copper	0.5–1.5 mg
Niacin	40 mg	Manganese	0.15–0.8 mg
Vitamin B_6	4 mg	Chromium	10–15 μg
Folacin	400 μg	Selenium	40–120 μg
Vitamin B_{12}	5μg	Sodium	40–180 mEq
Biotin	60 μg	Potassium	30–180 mEq
Pantothenic acid	15 mg	Chloride	40–300 mEq

Because of variability in intestinal absorption and utilization, these intravenous recommendations may differ from the oral requirements, particularly with regard to certain minerals and trace elements. However, since the absorption of protein, carbohydrate, and fat is nearly complete in normal individuals, the requirements for these nutrients are similar for the intravenous and oral routes. The amount of nutrition given is determined by the patient's present nutritional status, the overall goals of nutritional support (nutritional maintenance of the reasonably well-nourished individual or nutritional rehabilitation of the malnourished patient), and the patient's underlying disease state, which may affect both the nutritional requirements and the utilization of different nutrients.

Approximate Guidelines for Intake of Protein and Calories

1. Stable patient for nutritional maintenance
   Protein 1 g/kg per day
   Calories $1.2 \times$ BEE (or approximately 30 kcal/kg per day)
2. Nutritionally depleted patients for repletion
   Protein 1.5 g/kg per day
   Calories $1.5 \times$ BEE
3. Hypermetabolic patients*
   Protein 2 g/kg per day
   Calories $2 \times$ BEE

---

*Special caution must be taken in the presumed hypermetabolic patient to avoid overfeeding, particularly of carbohydrate. This can contribute to the patient's hypermetabolism without promoting improved protein anabolism. If possible, actual measurement of BEE should be performed.

A. Source of Calories

In choosing the ideal source of intravenous calories for the elderly, several factors suggest that a combination of carbohydrate and lipid, rather than either alone, is the most rational decision.

Factors Supporting Combined Protein/Lipid Calorie Sources

1. Glucose intolerance is common in the elderly; therefore, they may be prone to hyperglycemia resulting from high dextrose loads
2. High-dextrose infusions may promote salt and water retention, contributing to fluid overload or heart failure in the elderly patient with underlying heart disease and/or cardiac cachexia
3. Glucose infusions lead to increased carbon dioxide production, which will increase the ventilatory load; this may be particularly hazardous to elderly patient with pulmonary disease or during weaning from mechanical ventilators
4. Tubular reabsorption of glucose declines with age; this may lead to earlier glycosuria with concomitant salt and water loss in those elderly with hyperglycemia
5. Fatty infiltration of the liver during TPN infusions is related to continuous high-carbohydrate infusions
6. Most adults can metabolize 2.5 g of fat per kilogram of body weight per day, or 60 percent of total calories, without difficulty, even in the setting of advanced liver disease
7. Overall conservation of body protein, particularly in the long-term patient, is most likely similar whether fat or carbohydrate is used as the nonprotein calorie source

For these reasons, a balance of fat and carbohydrate may be advantageous.

B. Mode of Delivery of Parenteral Nutrition

Parenteral nutrition may be delivered by either a peripheral vein or a central venous catheter. In general, peripheral veins are useful in individuals requiring supplemental support in addition to marginal oral intake or in those expected to require short-term (less than 2 weeks) therapy. By peripheral i.v., approximately 1 g/kg per day of protein and 2000 kcal per day can be provided. Those requiring higher protein–calorie loads, particularly if fluids must be restricted, may need central line infusions. In addition, recurrent superficial vein phlebitis frequently limits extended use of this modality.

To provide prolonged intravenous nutritional support with hypertonic solutions, an indwelling central venous catheter inserted aseptically into the internal jugular or subclavian vein is required.

Representative Examples of Parenteral Nutrition Formulas for Central and Peripheral Line Administration

Nutrient	Peripheral vein	Central vein
Dextrose 10% (340 kcal/ liter)	1000 cc	—
Dextrose 50% (1700 kcal/liter)	—	1000 cc
8.5% Amino acids with electrolytes	1000 cc	1000 cc
Multiple trace element	5 cc	5 cc
MVI-12 (USV Laboratories)*	10 cc	10 cc
Lipid 10% (1.1 kcal/cc)	1000 cc	—
Lipid 20% (2.0 kcal/cc)	—	500 cc
Total volume	3015 cc	2515 cc
Total kcal	1780	3040
Total protein (g)	85	85
Osmolality (mOsm/ liter)	833	1840

*MVI-12 contains vitamin C, 100 mg; vitamin A, 3300 IU; vitamin D, 200 IU; thiamine, 3.0 mg; riboflavin, 3.6 mg; pyridoxine, 4.0; niacinamide, 40.0 mg; pantothenic acid, 15.0 mg; vitamin E, 10 IU; biotin, 60 μg; folic acid, 400 μg; vitamin B_{12}, 5 μg. Note: Vitamin K must be given separately, 5 mg IM weekly, if not on anticoagulants.

Approximate Electrolyte Concentration*

Nutrient	Peripheral vein	Central vein
Na (mEq/liter)	35	35
K (mEq/liter)	30	30
Cl (mEq/liter)	35	35

(continued)

Approximate Electrolyte Concentration*
(Continued)

Nutrient	Peripheral vein	Central vein
Acetate (mEq/liter)	65	65
PO$_4$ (mM)	15	15
Mg (mEq/liter)	5.0	5.0
Ca (mEq/liter)	4.8	4.8
Zn (mg/day)	4	4
Cu (mg/day)	1	1
Cr (μg/day)	10	10
Mn (mg/day)	0.8	0.8
Se (μg/day)	50	50

*Assuming a mixture of 500 cc dextrose plus 500 cc amino acids with electrolytes.

Special formulas are also available for the patient with cardiac, renal, or hepatic disease. In the cardiac patient, where fluid overload must be prevented, increasing the caloric density of the formula by using 70% dextrose will allow for greater calorie intake with less total volume. For the patient with liver disease and hepatic encephalopathy, electrolyte-free protein solutions rich in branched-chain amino acids (e.g., Hepatamine®) may promote positive nitrogen balance, reduce encephalopathy, and perhaps enhance survival in selected cirrhotic patients. For individuals with renal failure, electrolyte-free, low-total-protein, but high-essential-amino-acid formulas are available (e.g., Renamine®, Nephramine®) to enhance protein synthesis without promoting a marked rise in blood urea nitrogen (BUN). These special renal formulas are best suited for the predialysis patient with moderate renal insufficiency.

C. Complications of TPN

Because of potential complications during parenteral nutrition, careful clinical and laboratory monitoring of patients is advised. This should include a daily physical examination and other tests, as outlined below.

Suggested Laboratory and Clinical Monitoring of Hospitalized Elderly Patients Receiving Parenteral Nutritional Support

Parameter	Baseline	First week	Frequency Second week if stable	Thereafter if stable
CBC/differential	+	TIW*	BIW*	Weekly
PT, PTT	+	Weekly	Once	Weekly
Na, K, C1, CO_2	+	Daily	TIW	BIW
BUN, Glu	+	Daily	TIW	BIW
PO_4	+	Daily	TIW	BIW
Mg	+	TIW	BIW	Weekly
Bilirubin, alk phos, SGOT	+	TIW	BIW	Weekly
Albumin	+	TIW	BIW	Weekly
Calcium	+	TIW	BIW	Weekly
Zinc	+	Once	Once	q2wk
Urine glucose	+	q6h	q6h	q6h
Body weight	+	Daily	Daily	TIW
Serum iron/ TIBC	+	−	Once	Monthly
Triglycerides/ cholesterol	+	−	+	Monthly

*TIW, 3 times per week; BIW, 2 times per week.

Potential Complications of Parenteral Nutrition

1. Insertion of the catheter
   a. Pneumothorax
   b. Subclavian artery injury
   c. Brachial plexus injury
   d. Air embolism: may be reduced by Trendelenburg position and prior rehydration of the patient
2. Catheter, post-insertion
   a. Sepsis: strict aseptic technique with an occlusive dressing and specific nursing care protocols should keep sepsis rate to less than 7% even in the immunosuppressed patient
   b. Thrombosis of the subclavian vein
   c. Septic thrombosis
3. Metabolic complications
   a. Hyperglycemia
   b. Hypoglycemia

(continued)

Potential Complications of Parenteral Nutrition *(Continued)*

   c. Hyperglycemic hyperosmolar coma
   d. Hypokalemia
   e. Hypophosphatemia
   f. Hypomagnesemia
   g. Hyperchloremic metabolic acidosis
   h. Hypo- or hypernatremia
   i. Essential fatty acid deficiency in those not receiving
     lipid at least once per week
   j. Zinc deficiency
   k. Abnormal liver function tests, rarely causing
     jaundice

X. SUMMARY

Nutritional maintenance of both the healthy and the ill elderly patient represents a substantial clinical problem. Changes in nutrient requirements and utilization associated with normal aging, malnutrition, and age-prevalent disease combine to make advanced nutritional support with either enteral or parenteral techniques particularly challenging. Nutritional therapy, when necessary, can be safely administered with great benefit to properly selected and appropriately monitored geriatric patients.

SUGGESTED READING

Blackburn G. L., Thornton P. A.: Nutritional assessment of the hospitalized patient. *Med Clin North Am* 53:1103–1115, 1979.

Bowman B. B., Rosenberg I. H.: Assessment of the nutritional status of the elderly. *Am J Clin Nutr* 35:1142–1151, 1982.

Cahill G. F.: Starvation in man. *N Engl J Med* 282:668–675, 1970.

Cape R. D. T.: Geriatrics, in Schneider H. A., Anderson C. A., Coursin D. B. (eds): *Nutritional Support of Medical Practice.* Philadelphia, Harper & Row, 1983, pp 369–385.

Cataldi-Betcher E. L., Seltzer M. H., Slocum B. A., et al: Complications occurring during enteral nutrition support: A prospective study. *J Parent Ent Nutr* 7:546–552, 1983.

Dudrick S. J., MacFadyen B. U., Souchon E. A., et al: Parenteral nutrition techniques in cancer patients. *Cancer Res* 37:2440–2450, 1977.

Dudrick S. J., MacFadyen B. V., Van Buren C. T., et al: Parenteral hyperalimentation: Metabolic problems and solutions. *Ann Surg* 176:259–264, 1972.

Dworkin B.: Nutritional support of the geriatric patient, in Gambert S. (ed): *Contemporary Geriatric Medicine* (vol II). New York, Plenum Medical, 1986, pp. 375–412.

Harper A. E.: Recommended dietary allowances for the elderly. *Geriatrics* 33:73–80, 1978.

Harris J. A., Benedict F. G.: *A Biometric Study of Basal Metabolism in Man.* Washington, Carnegie Institution, 1919.

Holt P. R.: Intestinal absorption and malabsorption, in Texter E. C. (ed): *The Aging Gut.* New York, Mason USA, 1983, pp 33–56.

Jeejeebhoy K. N. (ed): *Total Parenteral Nutrition in the Hospital and at Home.* Boca Raton, CRC Press, 1983.

Koretz R. L., Meyer J. H.: Elemental diets: Facts and fantasies. *Gastroenterology* 78:393–410, 1980.

Lindeman R. D., Papper S.: Therapy of fluid and electrolyte disorders. *Ann Intern Med* 82:64–70, 1975.

Michel L., Serrano A., Malt R. A.: Nutritional support of hospitalized patients. *N Engl J Med* 304:1147–1152, 1981.

Mitchell C. O., Lipschitz D. A.: Detection of protein–calorie malnutrition in the elderly. *Am J Clin Nutr* 35:395–406, 1982.

Munro N. H.: Nutrition and aging. *Br Med Bull* 37:83–88, 1981.

Rivlin R.: Nutrition and aging: Some unanswered questions. *Am J Med* 741:337–340, 1981.

Roe D. A.: Interactions between drugs and nutrients. *Med Clin North Am* 63:985–1007, 1979.

Roe D. A.: *Geriatric Nutrition.* Englewood Cliffs, NJ, Prentice-Hall, 1983.

Shils M. E.: Enteral nutrition by tube. *Cancer Res* 37:2432–2439, 1977.

Shils M. E.: Parenteral nutrition, in Goodhart R. S., Shils M. E. (eds): *Modern Nutrition in Health and Disease* Philadelphia, Lea & Febiger, 1980, pp. 1125–1152.

Shils M. E., Bloch A. S., Chernoff R.: Liquid formulas for oral and tube feeding. *J Parent Ent Nutr* 1:89–96, 1977.

Shock N. W.: Physiologic aspects of aging. *J Am Diet Assoc* 56:491–496, 1970.

Stefee W. B.: Nutrition intervention in hospitalized geriatric patients. *Bull NY Acad Med* 56:564–574, 1980.

Watkin D. M.: Nutrition for the aging and the aged, in Goddhart R. S., Shils M. E. (eds): *Modern Nutrition in Health and Disease.* Philadelphia, Lea & Febiger, 1980, pp 781–813.

Young E. A.: Nutrition, aging, and the aged. *Med Clin North Am* 67:295–313, 1983.

12

Infectious Diseases in the Elderly

I. INTRODUCTION

Infections are a major cause of morbidity and mortality in the elderly population. The elderly have an increased susceptibility to infection not only because of longer and more frequent periods of hospitalization but also because of the impairment of host defense mechanisms associated with aging.

Age-Related Alterations in Host Defense Mechanisms

1. Decreased response to challenge by microorganisms
2. Decreased primary antibody response
3. A decrease in cell-mediated immunity

Elderly patients who are infected often have atypical presentations. Fever, an important clinical manifestation of infection, is frequently absent. The only complaints an infected elderly patient may have are anorexia, weakness, nausea, or fatigue. Some may only develop personality changes, delirium, or manifest an acute dementia. On this basis, it is important to consider infection as a cause of any recent or acute change in the elderly patient's general state of health.

II. BACTERIAL PNEUMONIA IN THE ELDERLY

A. General

1. Occurs more frequently than in other high-risk groups.
2. Mortality rates are high and depend on the etiologic agent.
3. Fourth most common (with influenza) and leading infectious cause of death.

B. Pathogenesis

1. Loss of pulmonary elastic tissue and weakening of respiratory muscles.
2. Impaired mucociliary transport.
3. Decreased forced vital capacity.
4. Coincident chronic lung disease.
5. Prior infection with influenza.

C. Clinical Features

1. Fever, cough, and purulent sputum may be absent.
2. Frequent insidious deterioration in general health.
3. Confusion is a common early sign.
4. May present as sudden deterioration of, or slow recovery from, a preexisting disease (i.e., congestive heart failure, stroke).
5. Physical examination frequently nonspecific.
6. Increased respiratory rate may be an early diagnostic clue.

D. Diagnosis

1. High index of suspicion.
2. Sputum gram stain and culture.
   a. Older patients are frequently dehydrated with a weak cough or produce a specimen composed of oropharyngeal secretions.
   b. Cultures should be restricted to specimens that show more than $25/100 \times$ field polymorphonuclear cells and sparse epithelial cells (fewer than $10/100 \times$ field).

3. Transtracheal aspiration is highly reliable but invasive and should only be performed by skilled persons in patients without contraindications (i.e., bleeding, hypoxemia, inability to cooperate).
4. Chest X ray may frequently show incomplete consolidation.
5. Blood cultures are frequently sterile but should be obtained prior to initiating antibiotic therapy.
6. Evaluate all pleural effusions.
7. Leukocytosis may be an early clue.
8. Obtain antibody titers (acute and convalescent) for influenza, *Mycoplasma pneumoniae*, or *Legionella pneumophila*.

E. Community-Acquired Pneumonia

1. Aspiration of bacteria in patient's oropharyngeal secretions.
2. Primarily caused by *S. pneumoniae* (40–60%).
3. Other etiologic agents include *H. influenzae* (2.5–20%), gram-negative bacilli (6–37%), *S. aureus* (2–10%), *L. pneumophila* (0–22.5%), and mouth anaerobes (unknown incidence).
4. Incidence of gram-negative pneumonia increases with age.
5. Higher mortality with infection by gram-negative bacilli and *S. aureus* (80%) than by *S. pneumoniae* (20%).
6. Initial antibiotic therapy should be based on sputum gram stain, the patient's drug allergy history, and adverse side effects of antibiotics.
7. Final treatment should be directed at the specific organism isolated from blood, pleural fluid, or sputum.
8. In cases in which no initial gram stain can be obtained, empiric antibiotic therapy should include:
   a. Ampicillin or cefamandole for the non-critically-ill patient.
   b. Ampicillin, nafcillin (or oxacillin), and an aminoglycoside or cefamandole and an aminoglycoside, or a third-generation cephalosporin and an aminoglycoside for the critically ill patient. In life-threatening situations when *L. pneumophila* is considered, add erythromycin to the regimen.
9. Therapy should last for 10–14 days, although the critically ill patient may need at least 3 weeks of antibiotics.

F. Nosocomial Pneumonia

1. Major threat associated with institutionalization of the elderly.
2. Comparable risk in hospitalized patients and those in nursing homes.
3. Risk factors include colonization with gram-negative bacilli, recent surgery, conditions predisposing to aspiration (i.e., depressed mental status), poor pulmonary toilet, recent respiratory instrumentation, and impaired immune function.
4. Primarily caused by gram-negative bacilli (40–60%).
5. Other etiologic agents include *S. aureus* (10–30%), mouth anaerobes (35%), and *S. pneumoniae* (10–20%). There is a high frequency of mixed infections.
6. Increasing prevalence of methicillin-resistant strains of *S. aureus*.
7. Additional associated microbial agents include influenza viruses and *Legionella pneumophila*. *Legionella pneumophila* commonly presents as an atypical pneumonia with constitutional symptoms (i.e., headache, myalgia, diarrhea) and extensive pulmonary involvement on chest X ray.
8. In cases in which no initial sputum gram stain can be obtained, empiric therapy includes an aminoglycoside combined with a cephalosporin, an aminoglycoside with clindamycin, or an aminoglycoside with a penicillinase-resistant penicillin. If *L. pneumophila* is considered, add erythromycin.

Bacterial Pneumonia in the Elderly

Etiologic agent	Associated factors	Drug(s) of first choice	Alternative drug(s)
Streptococcus pneumoniae	COPD Influenza Pulmonary tumor Chronic alcoholism Multiple myeloma	Penicillin G	Erythromycin Cephalosporins Chloramphenicol Vancomycin Clindamycin

(continued)

Bacterial Pneumonia in the Elderly (*Continued*)

Etiologic agent	Associated factors	Drug(s) of first choice	Alternative drug(s)
Hemophilus influenzae	COPD Pulmonary tumor Chronic alcoholism Influenza	Ampicillin* Cefamandole	Trimetho-prim–sulfa-methoxazole Cefuroxime Cefaclor Cefotaxime Tetracycline Chlorampheni-col
Mouth ana-erobes	Depressed mental status Dysphagia Pulmonary tumor Chronic alco-holism	Penicillin G Clindamycin	Cephalospo-rins Chlorampheni-col Erythromycin Tetracycline
Staphylo-coccus aureus (meth-icillin re-sistant)	Nosocomial Influenza	Nafcillin (oxacillin) Vancomycin	Cephalospo-rins Vancomycin Clindamycin Trimetho-prim–sulfa-methoxazole
Gram-nega-tive bacilli	Nosocomial Chronic alcoholism Granulocyto-penia	Aminoglycos-ide and a cephalo-sporin Aminoglycos-ide and a semisyn-thetic peni-cillin	
Legionella pneumo-phila	COPD Alcoholism Nosocomial Immunosup-pression	Erythromycin ± rifampin	Tetracycline

*Check incidence of ampicillin resistance in community.

III. URINARY TRACT INFECTION IN THE ELDERLY

A. General

1. Bacteriuria: $\geq$100,000 colony-forming units per milliliter of urine; may be symptomatic or asymptomatic.
2. Risk factors for urinary tract infection (UTI) include underlying disease(s), place of residence (i.e., nursing home), and genitourinary abnormalities.
3. Female to male ratio 2 : 1.
4. High incidence of associated bacteremia.
5. Most common cause of bacteremic sepsis in the elderly.

B. Clinical Features

1. Genitourinary complaints frequently absent.
2. Common complaints include nocturia, incontinence, stress incontinence, or urgency.
3. Atypical symptoms include lethargy, anorexia, dementia, or confusion.
4. Persistent or recurrent bacteriuria is common.

C. Diagnosis

1. Urinalysis (pyuria if >10 WBC/hpf).
2. Urine gram stain and culture.
3. Blood cultures to rule out urosepsis.
4. If bacteremia occurs with a community-acquired pyelonephritis, a radiographic evaluation of the genitourinary tract should be done.

D. Microbiology

1. Low isolation rate of *E. coli.*
2. Other gram-negative bacilli (i.e., *Klebsiella, Enterobacter, Pseudomonas*) as well as enterococci and *Staphylococcus epidermidis* predominate, especially in hospital-acquired and catheter-associated infections.

E. Therapy

1. Antibiotics based on urine gram stain and culture.
2. Increasing incidence of ampicillin-resistant *E. coli* in community- and hospital-acquired infections.
3. Treat uncomplicated UTIs for 7–14 days.
4. Treat recurrent UTIs with no urological abnormalities for 3–6 weeks.
5. Antibiotic treatment of asymptomatic bacteriuria in the elderly has a high failure rate.

F. Catheter-Associated Bacteriuria

1. Most common nosocomial infection.
2. Incidence of UTI is 5–10%/day of catheterization.
3. Risk factors include periurethral colonization with urinary tract pathogens.
4. Most patients with long-term indwelling catheters will have significant bacteriuria.
5. Closed catheter system decreases the incidence of bacteriuria.
6. Polymicrobial infection and resistant organisms are common.
7. Major risk is urosepsis with gram-negative bacilli, which carries a high mortality.
8. Bacteriology of urine samples obtained from indwelling catheters not changed for ≥30 days may not accurately reflect pathogens in bladder urine; the old catheter should be replaced, and urine obtained for bacteriology through a newly inserted catheter. Results of previous urine cultures do not necessarily correlate with current infecting organism(s).
9. Antibiotic use to eradicate bacteriuria in the long-term catheterized patient is ineffective. Only treat if a clinical change occurs indicating active infection. Initial antibiotic therapy should cover the prevalent organisms where the infection was acquired.
10. In general, aminoglycosides should be part of the initial antibiotic regimen pending urine culture results. Specific therapy should be based on urine gram stain and culture.
11. Treatment course should last 10–14 days.

IV. BACTERIAL MENINGITIS IN THE ELDERLY

A. General

1. Mortality rate high regardless of etiologic agent (>75%).
2. Increased incidence of neurological sequelae.
3. *Streptococcus pneumoniae* is the most common pathogen.
4. Other common etiologic agents include *L. monocytogenes, S. aureus,* and gram-negative bacilli.
5. *Hemophilus influenzae* and *N. meningitidis* are less common.

B. Clinical Features

1. Three types of presentation:
   a. Headache, confusion, lethargy of rapid onset.
   b. Progressive onset of meningismus.
   c. Primary respiratory disease 1–3 weeks prior to development of meningeal signs.
2. Delay in diagnosis common.
3. Patients frequently have an associated pneumonia.

C. Diagnosis

1. Immediate lumbar puncture is necessary unless patient has papilledema or focal neurological signs. In this case, computerized axial tomography (CT scan) prior to lumbar puncture should be done to rule out brain abscess or other mass lesions.
2. Lack of cellular response in the cerebrospinal fluid (CSF) is more likely to occur in the elderly patient with meningitis and is associated with a worse prognosis.
3. Usually the CSF has a predominance of polymorphonuclear leukocytes (≥90%).
4. Lymphocytes predominate in patients with meningitis caused by *L. monocytogenes.*
5. CSF glucose <40% of blood sugar is common.
6. Elevated CSF pressure and protein are usual.
7. Gram stain and culture of CSF are crucial to diagnosis.
8. Counterimmunoelectrophoresis to identify pneumococcal, meningococcal, type B *H. influenzae,* and *E. coli* K_1 antigens may be helpful.

9. *Limulus* lysate assay may be useful in diagnosing gram-negative meningitis.
10. Cultures of blood, urine, sputum, and wounds may be useful in determining the source of infection.

Bacterial Meningitis in the Elderly

Etiologic agent	Associated factors	Drug(s) of first choice	Alternative drug(s)
Streptococcus pneumoniae	Pneumonia Otitis media Skull fracture CSF rhinorrhea Mastoiditis	Penicillin G	Chloramphenicol Cefotaxime
Gram-negative bacilli	Neurosurgery Head trauma Urinary tract infection or manipulation Pneumonia ?Decubitus ulcers	Cefotaxime Aminoglycoside (Carbenicillin/ticarcillin and aminoglycoside if CSF grows *Pseudomonas*)	Chloramphenicol Trimethoprim–sulfamethoxazole Intrathecal aminoglycosides
Listeria monocytogenes	Lymphoproliferative disease No underlying disease Steroids	Penicillin G Ampicillin ± aminoglycoside	Chloramphenicol ?Trimethoprim–sulfamethoxazole
Staphylococcus aureus	Bacterial endocarditis Neurosurgery Pneumonia	Nafcillin (oxacillin)	Vancomycin
Neisseria meningitidis	Pharyngitis, epidemic	Penicillin G	Chloramphenicol Cefotaxime Cefuroxime
Hemophilus influenzae	Otitis media Pneumonia Immunodeficiency	Ampicillin and chloramphenicol (May switch to ampicillin alone if β-lactamase negative)	Cefotaxime Cefuroxime Trimethoprim–sulfamethoxazole

D. Management

1. Rapid initiation of antibiotics is essential.
2. Delay in diagnosis leads to higher mortality.
3. Assess hemodynamic status (i.e., shock).
4. Treat associated seizures.
5. Repeat lumber puncture in patients with relapsing or prolonged fever, no clinical improvement, or worsening clinical status.
6. Rifampin prophylaxis is recommended for intimate contacts of patients with meningococcal meningitis and possibly of those with *H. influenzae* meningitis.

V. ABDOMINAL INFECTIONS IN THE ELDERLY

A. General

1. Increased morbidity and mortality in the elderly.
2. Atypical presentation common with a paucity of history and clinical findings.

B. Biliary Tract Infection

1. Increased frequency of cholecystitis and cholangitis.
2. Fever, leukocytosis, and abdominal pain or tenderness may be absent.
3. Common microbes include *E. coli, Klebsiella,* enterococcus, and *Bacteroides fragilis* (increased incidence of anaerobes in the elderly).
4. Increased septic complications.
5. Acalculous cholecystitis common.
6. Empiric antibiotic therapy includes ampicillin or penicillin, an aminoglycoside, and clindamycin or metronidazole.
7. Early surgery recommended.

C. Diverticulitis

1. Occurs more commonly with advanced age.
2. Nonspecific complaints common.

3. Left lower quadrant tenderness and palpable mass may be present.
4. Polymicrobial bacteremia occurs.
5. Empiric antibiotic therapy in severely ill patients should include an aminoglycoside and clindamycin or metronidazole.
6. Surgery required in 20–30% of hospitalized patients.

D. Appendicitis

1. Increased mortality rate of 2–14% in patients >60 years.
2. Delay in diagnosis common.
3. Atypical presentations are common, but the majority complain of abdominal pain, nausea, anorexia, or a change in bowel habits.
4. Leukocytosis occurs in 75–90% of cases.
5. Advanced pathology of appendix at surgery (i.e., gangrene or perforation) is common.
6. Prompt surgery is necessary.
7. Empiric antibiotic therapy in complicated cases includes ampicillin or penicillin, an aminoglycoside, and clindamycin or metronidazole.

E. Intraabdominal Abscess

1. Mortality approximately 35%.
2. Most patients are seriously ill with fever, chills, and abdominal pain.
3. Palpable abdominal mass is present in <10% of cases.
4. The majority have leukocytosis.
5. Diagnosis is often delayed.
6. The CT scan is the best diagnostic procedure.
7. Mixed infection, including aerobes and anaerobes, is the rule.
8. Primary therapy should include surgical drainage.
9. Empiric antibiotic therapy includes an aminoglycoside and clindamycin or metronidazole with or without ampicillin or penicillin.

VI. TUBERCULOSIS IN THE ELDERLY

A. General

1. Increasing proportion of all forms of reactivation tuberculosis occurs in patients >60 years of age.
2. Atypical presentations and chronic illness often obscure the diagnosis.

B. Clinical Presentation

1. Although weight loss, cough, and weakness are the most frequent symptoms, they are often mild and chronic.
2. Alcohol abuse, previous gastrectomy, and silicosis are associated with an increased risk of tuberculosis.
3. There is a frequent occurrence of other medical diseases (i.e., COPD, heart disease, and lung cancer).
4. Physical examination is often nonrevealing.
5. Fever is the exception, not the rule.

C. Diagnosis

1. Tuberculin skin test and anergy screen.
2. Unusual radiographic findings in one-third of patients.
3. Sputum for acid-fast stain and culture.
4. Definitive diagnosis rests on identification of the etiologic agent (*M. tuberculosis*) from sputum, body fluids, or tissue.
5. There is a high incidence of pleural disease caused by *M. tuberculosis*.
6. Other diagnostic methods include gastric lavage, bronchoscopy, and pleural biopsy (if pleural effusion is present).

D. Extrapulmonary Tuberculosis

1. Highest rates occur in the elderly population.
2. Renal involvement is common; urine should be cultured (early morning specimen).
3. Skeletal tuberculosis should be considered in any elderly

     patient with unexplained back pain or unexplained unifocal
     inflammation or destruction of a bone or joint.
4. Tuberculous meningitis has a second peak incidence in
     patients >65 years of age and has a subacute presentation.
5. The elderly constitute the majority of cases with late
     miliary tuberculosis.
6. Liver biopsy is positive in 90% of patients with miliary
     tuberculosis, whereas bone marrow biopsies are only
     positive in approximately 33% of cases.

E. Therapy

1. First-line drugs for untreated pulmonary tuberculosis in-
     clude isoniazid and rifampin.
2. Choice of antibiotics and duration of therapy (i.e., 9
     months versus 18 to 24 months) are dictated by the
     patient's previous therapy, coexisting illnesses, need for
     other medications, and ability to tolerate the drugs.
3. If prior therapy has been given, use at least two drugs to
     which the patient has not been previously exposed.
4. Use caution when using isoniazid and rifampin in a patient
     with preexisting liver disease.
5. Renal insufficiency may result in toxic levels of
     ethambutol.

VII. INFECTIVE ENDOCARDITIS IN THE ELDERLY

A. General

1. Increasing incidence in the elderly population.
2. Male predisposition.
3. Approximately 50% of patients have no prior history of
     heart disease.

B. Predisposing Factors

1. Arteriosclerotic heart disease is present in 33% of cases.
2. Prior dental procedure in 15–20% of patients with nonen-
     terococcal streptococcal endocarditis.

3. Prior genitourinary tract procedure in 42% of patients with enteroccocal endocarditis.
4. Prior procedure (9%) and prior underlying infection (35%) found in patients with staphylococcal endocarditis.
5. Intravenous lines are the most likely source of infection in hospital-acquired endocarditis.
6. *Streptococcus bovis* endocarditis is associated with underlying gastrointestinal lesions, especially colon carcinoma.
7. Other sources of infection include decubitus ulcers, gastrointestinal procedures, and systemic infection.
8. No apparent source of infection is found in 50% of patients.

C. Clinical Features

1. Atypical presentations are common and frequently lead to a delay in diagnosis.
2. Fever and murmur are often absent.
3. Neurological symptoms and signs are common; their presence increases the mortality rate to 75%.
4. Urinary symptoms occur frequently.
5. Congestive heart failure (53%) and azotemia (44%) often occur.
6. Cutaneous lesions (petechiae, splinter hemorrhages, Osler's nodes, Janeway lesions, clubbing) occur in 33% of cases.
7. Psychiatric symptoms (depression, confusion, paranoia) are common and may be the only presenting signs.

D. Microbiology

1. Streptococci, including enterococci and *S. bovis*, cause 25–70% of cases.
2. Staphylococci, especially *S. aureus*, cause 20–30% of cases and have a high mortality rate.
3. Gram-negative organisms causing endocarditis are more common in the elderly because of the high incidence of nosocomial, genitourinary, and gastrointestinal tract infections.
4. *Bacteroides* species rarely cause endocarditis; endocarditis

caused by these organisms, however, is more common in
the elderly population.

5. Culture-negative endocarditis accounts for 10–20% of
cases.

E. Laboratory Findings

1. If positive, the first two sets of blood cultures will yield the
organism in greater than 90% of patients.
2. Other abnormalities include anemia, microscopic hematuria,
abnormal renal function, and elevated erythrocyte sedimentation rate.
3. Leukocytosis may be absent.
4. Rheumatoid factor is more commonly found in elderly
patients with endocarditis.

F. Treatment

1. The treatment of choice for enterococcal endocarditis is
penicillin G or ampicillin combined with an aminoglycoside
(i.e., gentamicin).
2. Penicillin-sensitive streptococci (S. viridans and S. bovis)
are effectively treated with high doses of penicillin in
patients over 65 years of age. The combination of penicillin
G and an aminoglycoside may be given for a shorter time
but entails the risk of aminoglycoside toxicity in the elderly
patient.
3. Staphylococcal endocarditis is treated with penicillinase-
resistant penicillins, often combined with aminoglycosides
or rifampin. Vancomycin is the treatment of choice for
methicillin-resistant staphylococci or in the elderly patient
who is allergic to penicillin.
4. Empiric antibiotic therapy includes penicillin G or ampicillin plus a penicillinase-resistant penicillin and an aminoglycoside or vancomycin and an aminoglycoside.
5. Serum drug levels and renal function must be closely
monitored to avoid toxicity.
6. Serum bactericidal titers, when greater than or equal to
1 : 8, correlate with higher cure rates.

7. A poorer prognosis occurs in the elderly because of common delays in diagnosis, more frequent infections with *S.
aureus*, and a higher incidence of underlying cardiovascular
disease.
8. Mortality rates approach 40–60%.

VIII. ANTIBIOTICS

It is beyond the scope of this chapter to give dosages of
antibiotics for each clinical situation. Exact dosages required vary
with disease and individual patient profile. The following sections
include guidelines for prescribing antibiotics in the elderly patient.
The package insert for each drug should be consulted for use and
dosage as approved by the Food and Drug Administration. Other
reference sources for dosage of antibiotics include the *Physicians'
Desk Reference* and *The Medical Letter Handbook of Antimicrobial
Therapy*.

A. General

1. Absorption, distribution, metabolism, and excretion of antibiotics are all altered in the elderly. Thus, there is an
increased incidence of adverse side effects in this
population.
2. Renal insufficiency, which occurs with aging, has a major
influence on the pharmacokinetics of many antibiotics.
3. Creatinine clearance (Cr_{cl}) steadily decreases with age and
may be estimated by using the following formula:

Cr_{cl} (ml/min) = (140 − age) × weight (kg)/(serum
creatinine × 72)

This is to be multipled by 0.85 for women.

B. Penicillins

1. High dosages are necessary in the elderly patient with
meningitis or infective endocarditis.
2. Metabolic and neurological toxicity may occur in the

elderly patient with renal dysfunction if dosages are not properly adjusted.

3. Many penicillins contain a large amount of sodium and must be given cautiously to the elderly patient (i.e., carbenicillin, ticarcillin, oxacillin).

4. Hypokalemic alkalosis may occur with carbenicillin, ticarcillin, and sodium penicillin.

5. Carbenicillin inhibits platelet aggregation and fibrin formation and increases plasma antithrombin III activity. It should be given cautiously in patients who are uremic or are receiving other medications that affect coagulation.

6. Ampicillin is frequently associated with diarrhea and colitis; transient diarrhea is a common side effect of many antibiotics.

7. Methicillin is associated with interstitial nephritis more commonly than nafcillin or oxacillin.

8. Penicillin is excreted by the kidney; its half-life is increased in the elderly because of its reduced excretion.

9. Nafcillin, oxacillin, cloxacillin, and dicloxacillin are only slightly affected by renal insufficiency.

10. Ampicillin does not require major alteration in dosing for patients with renal insufficiency unless the Cr_{cl} is less than 30 ml/min. The pharmocokinetics of mezlocillin, pipercillin, and azlocillin are similar to those of ampicillin.

11. The serum half-life of both carbenicillin and ticarcillin is markedly prolonged in patients with anuria; dosages must be adjusted for renal insufficiency.

C. Cephalosporins

1. Renal damage is uncommon but well documented.

2. Interstitial nephritis may occur with cephalothin and is not dose related.

3. Most parenteral forms contain a significant amount of sodium; all oral forms are sodium-free.

4. Coagulation defects have been reported with high doses of cephalothin.

5. Moxalactam causes significant bleeding and interferes with hemostasis through either hypothrombinemia, platelet dysfunction, or immune-mediated thrombocytopenia. Patients

should be given 10 mg of vitamin K per week prophylactically while receiving this antibiotic. Bleeding times should be monitored in patients who receive more than 4 g of moxalactam per day for more than 3 days.

6. Cefoperazone can cause bleeding as a result of hypoprothrombinemia; patients receiving the drug should receive 10 mg of vitamin K per week prophylactically.

7. Neurotoxic reactions may also occur with high doses of the cephalosporins.

8. Cephalothin increases aminoglycoside nephrotoxicity (may be true of other cephalosporins as well).

9. Cefamandole, cefoperazone, and moxalactam produce a disulfiramlike reaction in patients drinking alcohol.

10. All doses of cephalosporins must be adjusted in patients with renal insufficiency except the third-generation cephalosporin cefoperazone, which is excreted primarily by the liver.

11. The third-generation cephalosporins cefotaxime and moxalactam as well as the second-generation cephalosporin cefuroxime cross the blood–brain barrier.

D. Aminoglycosides

1. The margin between therapeutic and toxic doses is narrow, especially in the elderly patient.

2. A borderline-normal serum creatinine in an elderly patient with little muscle mass may indicate significant renal insufficiency.

3. Serum levels of aminoglycosides and serum creatinine should always be closely monitored in the elderly patient.

4. Ototoxicity is more common in elderly patients with renal insufficiency or preexisting hearing loss.

5. Auditory nerve toxicity is more common when aminoglycosides are used with diuretics or when a patient is in renal failure. Stopping the drug may prevent further damage.

6. Aminoglycoside nephrotoxicity may be prolonged and may rarely be irreversible.

7. Neuromuscular paralysis is a rare but potentially serious side effect and may be treated by prompt administration of calcium.

8. The loading dose for patients with serious infection should be 2 mg/kg for gentamicin or tobramycin and 7.5 mg/kg for amikacin. The loading dose is the same in patients with renal insufficiency or normal renal function.
9. The maintenance dose of aminoglycosides should be based on the Cr_{cl}. Peak and trough antibiotic levels should be measured every few days to determine proper dosage.

E. Clindamycin

1. Pseudomembranous colitis occurs more frequently in elderly patients. Treatment consists of stopping the drug and initiating oral vancomycin (nonabsorbable).
2. The drug is metabolized in the liver; hepatic dysfunction may produce a fivefold increase in serum half-life.
3. Dosage should be reduced in patients with liver disease.

F. Chloramphenicol

1. Crosses the blood–brain barrier in the presence of inflamed and normal meninges.
2. Dose-related anemia results from reversible bone marrow depression.
3. Irreversible aplastic anemia is an idiosyncratic reaction.
4. Metabolized primarily in the liver; dosage must be adjusted in hepatic insufficiency.

G. Trimethoprim–Sulfamethoxazole

1. Adverse reactions are more frequent in females but not related to age.
2. Nausea, vomiting, and rarely diarrhea may occur.
3. If skin rash occurs, the drug should be stopped immediately.
4. Edlerly patients with low serum folate levels (i.e., malnutrition, malabsorption, alcoholism) are at risk of developing hematologic toxicity; folic acid administration is indicated in this situation.

5. Elevation of serum creatinine may occur.
6. Dosages should be adjusted for renal insufficiency when the Cr_{cl} is less than 50 ml/min.

H. Erythromycin

1. Adverse reactions uncommon.
2. Cholestatic jaundice may follow administration of the estolate form in 10% of treated patients and is reversible on stopping the drug.
3. Dose-related epigastric distress, nausea, and vomiting are common.
4. Excretion is primarily in the bile and feces.
5. Dosage change in renal failure is not recommended unless the Cr_{cl} is $\leq$10% of normal.
6. Elderly patients with slow gastrointestional motility may have lower serum levels because of acid degradation.
7. The estolate salt should be avoided in the geriatric patient because of the risk of hepatotoxicity.

I. Tetracyclines

1. Rarely the antibiotic of choice in the elderly.
2. Diarrhea common.
3. Demeclocycline may cause phototoxic dermatitis.
4. Vestibular reactions may occur, especially with minocycline.
5. May produce renal failure rapidly in patients with preexisting renal insufficiency.
6. Hepatotoxicity may occur following parenteral doses greater than 2 g per day or with normal doses in patients with renal insufficiency.
7. Primarily excreted by the kidney.

J. Vancomycin

1. "Red-neck syndrome" (tingling and flushing of face, neck, and thorax) may occur with rapid infusion of the drug.
2. Rash occurs in 4–5% of patients.

3. Auditory nerve damage and hearing loss most important adverse reactions.
4. Nephrotoxicity uncommon, but high doses should be avoided parenterally, and serum levels must be closely monitored.
5. High incidence of nephrotoxicity when both vancomycin and an aminoglycoside are given.
6. Excreted by the kidney; dosages must be adjusted in renal insufficiency.

K. Metronidazole

1. Major adverse reactions include nausea, vomiting, abdominal pain, and metallic taste.
2. Disulfiramlike reaction occurs in 25% of patients.
3. Parenteral administration is associated with phlebitis in 6% of patients.
4. Metabolized in the liver; dosage adjustment necessary in hepatic insufficiency.

L. Amphotericin B

1. Fever, chills, nausea, and headache may occur within minutes or hours of administration.
2. Phlebitis occurs in 70% of patients.
3. Renal toxicity is the most serious side effect.
4. Hypokalemia, hypocalcemia, and hypomagnesemia may occur.
5. Initial dosage should not be reduced in patients with renal failure.
6. If renal function deteriorates, amphotericin B should be changed to an every-other-day schedule or discontinued until the serum creatinine returns approximately to normal.

M. Comment

Antibiotics may cause serious problems if not given appropriately. The dose of a drug should be based on the patient's hepatic and renal function and the pharmacokinetic properties of each drug.

Renal and liver function tests as well as serum drug levels must be followed to minimize toxicity.

SUGGESTED READING

Bentley D. W.: Bacterial pneumonia in the elderly: Clinical features, diagnosis, etiology, and treatment. *Gerontology* 30:297–307, 1984.

Berger S. A.: Special considerations in the use of the new cephalosporins in the elderly. *Geriatr Med Today* 4:35–43, 1985.

Berk S. L.: Bacterial meningitis, in Gleckman R. A., Gantz N. M. (eds): *Infections in the Elderly.* Boston, Little, Brown, 1983, pp 235–263.

Cantrell M., Yoshikawa T. T.: Infective endocarditis in the aging patient. *Gerontology* 30:316–326, 1984.

DeTorres O. H., Marr F. N.: Antimicrobial agents, in Gleckman R. A., Gantz N. M. (eds): *Infections in the Elderly,* Boston, Little, Brown, 1983, pp 13–52.

Esposito A. L., Pennington J. E.: The pathogenesis of bacterial pneumonia, in Gleckman R. A., Gantz N. M. (eds): *Infections in the Elderly.* Boston, Little, Brown, 1983, pp 53–62.

Gantz N. M.: Infective endocarditis, in Gleckman R. A., Gantz N. M. (eds): *Infections in the Elderly.* Boston, Little, Brown, 1983, pp 217–234.

Gaynes R. P., Weinstein R. A., Chamberlin W., et al: Antibiotic-resistant flora in nursing home patients admitted to the hospital. *Arch Intern Med* 145:1804–1807, 1985.

Grahn D., Norman D. C., White M. L., et al: Validity of urinary catheter specimen for diagnosis of urinary tract infection in the elderly. *Arch Intern Med* 145:1858–1860, 1985.

Muder R. R., Yu V. L.: Legionnaires' disease: An emerging problem for geriatric patients. *Geriatr Med Today* 4:63–75, 1985.

Nagamu P., Yoshikawa T. T.: Aging and tuberculosis. *Gerontology* 30:308–315, 1984.

Newton J. E., Wilczynski P. J. G.: Meningitis in the elderly. *Lancet* 2:157, 1979.

Norman D. C., Yoshikawa T. T.: Intraabdominal infection: Diagnosis and treatment in the elderly patient. *Gerontology* 30:327–338, 1984.

Phair J. P.: 1983, Host defense in the aged, in Gleckman R. A., Gantz N. M. (eds): *Infections in The Elderly.* Boston, Little, Brown, 1983, pp 1–12.

Stead W. W.: Tuberculosis among elderly persons: An outbreak in a nursing home. *Ann Intern Med* 94:606–610, 1981.

Warren J. W.: Nosocomial urinary tract infections, in Gleckman R. A., Gantz N. M. (eds): *Infections in The Elderly.* Boston, Little, Brown, 1983, pp 283–318.

Yoshikawa T. T.: Unique aspects of urinary tract infection in the geriatric population. *Gerontology* 30:339–344, 1984.

13

Selected Neurological Problems in the Elderly

I. INTRODUCTION

Neurological disorders are not uncommon in the elderly. Fortunately, recent advances in diagnosis and treatment have reduced both morbidity and mortality associated with these problems. Alterations in the peripheral and central nervous system, however, remain a leading cause of institutionalization and altered life style in the elderly.

Effect of Age on Neurological Function

1. Loss of neurons
2. Decreased neuronal dendritic processes
3. Widening of brain sulci
4. Decrease in brain weight
5. Reduction in white matter
6. Neurofibrillary tangles and plaques in hippocampus
7. Decrease nerve conduction velocity
8. Increased reflex response time

"Normal" Age-Related Findings Commonly Noted on Neurological Examination

1. Irregular pupils
2. Sluggish light-reflex reaction
3. Poor pupillary reaction
4. Limited upward gaze
5. Slowed smooth-pursuit eye movements
6. Cogwheel tracking movements may be noted
7. Decreased muscle mass
8. Diminished reflex responses
9. Diminished ankle jerks
10. Absent superficial abdominal reflexes
11. Unresponsive plantar reflex
12. Palmomental and snout reflex may be present
13. Decreased sense of pain, touch, and vibration
14. Senile tremor

II. GAIT ABNORMALITIES IN THE ELDERLY

Gait disturbances are common among the elderly and can be caused by a variety of musculoskeletal and neurological disorders. Both normal age-related changes and age-prevalent illness must be considered in the differential diagnosis.

Common Causes of Abnormal Gait in the Elderly

Neurological

1. Impaired proprioception
2. Cerebellar dysfunction
3. Vestibular system abnormality
4. Normal-pressure hydrocephalus
5. Central nervous system lesion
6. Peripheral nerve disorder

Musculoskeletal

1. Ankylosis
2. Arthritis
3. Asymmetric limb length
4. Foot pathology
5. Shortened Achilles' tendon
6. Intermittent claudication
7. Bony alteration of spine and/or pelvis

Factors Suggesting Presence of a Neurogenic Gait Abnormality

1. Broad base
2. Unusual lifting of leg or foot
3. Dragging of foot or leg
4. Slow ambulation
5. Deviation of the trunk from side to side
6. Audible sound with gait: scraping, dragging, flopping, stamping

Classification of Abnormal Gaits

1. Sensory ataxic
2. Cerebellar ataxic
3. Spastic
4. Spastic ataxic
5. Parkinsonian
6. Steppage
7. Waddling
8. Choreoathetoid
9. Dystonic
10. Staggering
11. Toppling
12. Senile (*marche à petits pas*)
13. Vestibular
14. Frontal lobe

Characteristics of Gait Disorders

1. Sensory Ataxic

Characteristic Gait

Wide base; high-stepping; stamping

Clinical Findings

Romberg (+); eye closure worsens gait

Site of Pathology

Peripheral nerve, dorsal root, or posterior columns of
   spinal cord

Associated Illnesses

Tabes dorsalis
Friedreich's ataxia
Subacute combined degeneration (vitamin B_{12} deficiency)
Syphilitic meningomyelitis

(*continued*)

Characteristics of Gait Disorders (*Continued*)

1. Sensory Ataxic (*Continued*)

Associated Illnesses (*Continued*)

Chronic sensory polyneuropathies
Multiple sclerosis
Spinal cord compression

2. Cerebellar Ataxic Gait

Characteristic Gait

Wide-based; irregular; unsteady; staggering; falling to
    one side; short, irregular steps

Clinical Findings

Increased unsteadiness on quick turns, stops, or on get-
    ting up from a sitting position
Poor heel or toe walking
Romberg (+); can progress to inability to stand unaided

Site of Pathology

Bilateral or midline cerebellar dysfunction

Associated Illnesses

Degenerative disease (nutrition; alcohol)
Carcinoma (medulloblastoma, hemangioblastoma,
    metastasis)
Lesions to contralateral thalamic or frontal area; infarc-
    tion; hemorrhage; vascular disease

3. Spastic

a. Hemiplegic
b. Spastic paraparesis

Characteristic Gait

a. Leg held stiff; little flexion at hip or knees; outward leg
    rotation (circumduction)
b. Legs moved slowly and stiffly; limbs extended or
    slightly bent at the knees, tending to cross (scissors
    gait)

Clinical Findings

Associated neurological abnormalities common

(*continued*)

Characteristics of Gait Disorders (*Continued*)

3. Spastic (*Continued*)

Site of Pathology

Cerebral dysfunction, i.e., any hemiplegia, hemiparesis, paraparesis, or diffuse brain damage; spinal cord disorder with pyramidal-tract involvement

Associated Illnesses

a. Contralateral cerebral infarction; mass lesions; cerebral trauma; unilateral corticospinal-pathway dysfunction
b. Chronic spinal-cord compression; pernicious anemia; parasagittal mass lesion; multiple sclerosis; cerebral anoxia or injury

4. Spastic Ataxic

Characteristic Gait

Wide-based; unsteady; scissors gait

Clinical Findings

Romberg (+)

Site of Pathology

Combined involvement of the pyramidal tract and the cerebellar systems or posterior column

Associated Illnesses

Pernicious anemia; multiple sclerosis, vascular malformations; arachnoiditis; neoplasms

5. Parkinsonism

Characteristic Gait

Short, shuffling steps; stooped posture

Clinical Findings

Often fall forward when walking (propulsion); demonstrate increased speed of walking (festination)

Site of Pathology

Dopaminergic neurons in corpus stratum

(*continued*)

Characteristics of Gait Disorders (*Continued*)

5. Parkinsonism (*Continued*)
Associated Illness

Parkinson's disease; drug-induced parkinsonism; Huntington's disease; calcification of basal ganglia; manganese and carbon monoxide poisoning; striatonigral degeneration

6. Steppage Gait
Characteristic Gait

Feet lifted high; toes strike floor first

Clinical Findings

Slapping sound with gait; foot drop

Site of Pathology

Weakness or paralysis of the pretibial and peroneal muscles; abnormal peripheral nerves of the legs or motor neurons in the spinal cord; lesions of L_4–S_1 or cauda equina

Associated Illnesses

Paralysis of the peroneal or sciatic nerve; cauda equina lesions; Charcot–Marie–Tooth disease; progressive spinal muscular atrophy; ruptured intervertebral disk

7. Waddling Gait
Characteristic Gait

Weight placed alternately on each leg

Clinical Findings

Related to underlying pathology

Site of Pathology

Gluteal muscle weakness

Associated Illnesses

Muscular dystrophy; chronic spinal muscular atrophy; proximal myopathies (polymyositis); proximal neuropathies

(*continued*)

Characteristics of Gait Disorders (*Continued*)

8. Choreoathetoid

Characteristic Gait

Continuous, irregular movements of face, neck, hands, large proximal joints, and trunk; varying body contour with each step

Clinical Findings

Head jerking; grimacing; trunk and limb twisting; dancing characteristic

Site of Pathology

CNS dysfunction

Associated Illnesses

Huntington's chorea; degenerative, toxic, metabolic, vascular, neoplastic, or infectious disease of the CNS

9. Dystonic

Characteristic Gait

Bizarre; bowing; stepping; prominant limb flexion with prominent buttocks (lumbar lordosis)

Clinical Findings

Lordosis or scoliosis; facial grimacing; torticollis; tortipelvis; lower extremity flexion; rare in late life

Site of Pathology

CNS dysfunction

Associated Illnesses

Wilson's disease; Huntington's chorea; olivopontocerebellar atrophy; head trauma; cerebral injury; brain tumor; toxin

10. Staggering

Characteristic Gait

Irregular steeps; lurching; narrow base

Clinical Findings

Balance good; Romberg (−)

(*continued*)

Characteristics of Gait Disorders (*Continued*)

10. Staggering (*Continued*)

Site of Pathology

CNS dysfunction

Associated Illnesses

Alcoholic or barbiturate intoxication; bilaterial labyrinthine–vestibular dysfunction

11. Toppling

Characteristic Gait

Uncertain, hesitant steps

Clinical Findings

Falls common

Site of Pathology

Brainstem lesion

Associated Illnesses

Brainstem lesions; lateral medullary syndrome; progressive supranuclear palsy

12. Senile

Characteristic Gait

Short, uncertain steps; flexed posture

Clinical Findings

Diminished ability to respond to change; ungraceful

Site of Pathology

Presumed frontal lobe dysfunction

Associated Illnesses

Alzheimer's disease; cerebral lesions

13. Vestibular (Ataxic)

Characteristic Gait

Drifting noted to side of abnormality with prompt
correction

(*continued*)

Characteristics of Gait Disorders (*Continued*)

13. Vestibular (Ataxic) (*Continued*)

Clinical Findings

Unsteadiness, made worse with rapid movement; staggering when eyes closed; Romberg (+)

Site of Pathology

Vestibular system

Associated Illnesses

Chronic unilateral vestibular dysfunction; drug toxicity (streptomycin, neomycin, kanamycin); labyrinthine disease; cerebellopontine angle tumor

14. Frontal Lobe

Characteristic Gait

Wide-based; flexed posture; shuffling; hesitant

Clinical Findings

Halting noted; slow, awkward leg movement; difficulty sitting or standing

Site of Pathology

Cerebral dysfunction

Associated Illnesses

Alzheimer's disease; frontal lobe lesions; normal-pressure hydrocephalus

III. PARKINSONISM

A disease largely affecting those over 50 years of age, parkinsonism is characterized by tremor, rigidity, and akinesia. Although a variety of disorders and conditions that affect the central nervous system can result in this syndrome, idiopathic parkinsonism is responsible for the majority of cases.

The major biochemical abnormality in patients with parkinsonism is a loss or inhibition of activity of the neurotransmitter dopamine in the corpus striatum, resulting in an abnormal balance between dopamine and acetylcholine.

Forms of Parkinsonism

Primary (idiopathic)
  Parkinson's disease
Secondary (symptomatic)
  Infections: postviral encephalitis
  Arteriosclerotic
  Drug-induced
  Toxins: carbon monoxide, manganese, MPTP
  Metabolic: anoxia; parathyroid dysfunction
  Misc.: tumors; head trauma; degenerative; striatonigral
    degeneration; supranuclear palsy; olivopontocerebellar
    atrophy; autonomic degeneration (Shy–Drager; multi-
    system atrophy)

Treatment is directed towards correcting the balance between dopamine and acetylcholine in the CNS. In addition, physical therapy and psychological counseling can increase functioning. All patients are advised to establish a daily program of exercise. Physiotherapy is directed toward the maintenance of joint mobility, correction and prevention of postural abnormalities of limbs and trunk, and maintenance of normal gait.

Passive stretching of limbs, muscle massage, and gait training are useful adjuncts in therapy. (See table on p. 333.)

Patients with well-established signs and symptoms of parkinsonism who experience difficulties in conducting their activities of daily living and in their business and/or social obligations require pharmocological treatment.

Initially, they are best treated with agents capable of replenishing striatal dopamine; as the disease progresses, agents can be added to modify dopamine catabolism or mimic its action.

Carbidopa/Levodopa (Sinemet®) Levodopa is converted in the brain to dopamine. By combining it with carbidopa, a peripheral decarboxylase inhibitor, higher dopamine concentrations in the central nervous system can be obtained. Treatment should be started with low doses, 10/100 or 25/100 given t.i.d. and increased by one tablet every 3 days. The 25/100 tablet is usually more efficacious. In general, patients begin to show signs of improvement when a total daily dose of 125 mg/500 mg is reached. Further increases in dosage can be made at monthly intervals as needed. The optimal dose is that which improves function yet has minimal side effects. The usual therapeutic dosage range is 70 mg/700 mg

Medications Commonly Used in the Treatment of Parkinsonism

Medication	Trade name	Dosage form	Daily dosage range	Mechanism of action	Common side effects
Anticholinergics					
Benztropine mesylate	Cogentin	1-, 2-mg tablets	1–6 mg	Counteracts acetylcholine	Anticholinergic effects
Trihexyphenidyl HCl	Artane	Parenteral solution (1 mg/2 cc vial)	1 mg		
Antihistamines					
Diphenhydramine	Benadryl	25-, 50-mg capsules	25–200 mg	Counteracts muscarinic effects of acetylcholine	Sedation; anticholinergic effects
Dopamine-releasing or uptake inhibitors					
Amantadine HCl	Symmetrel	100-mg capsules	100–400 mg	Not fully understood; releases dopamine	Confusional state; edema; livedo reticularis on limbs
Amitriptyline HCl	Elavil	10-, 25-, 50-mg tablets	30–150 mg	Anticholinergic action; blocks reuptake and storage of dopamine	Anticholinergic effects
Imipramine	Tofranil				
Dopaminergics					
Carbidopa + levodopa	Sinemet	10/100-, 25/100-, 25/250-mg tablets	100/500–150/1500 mg.	Increases dopamine	Confusion; toxic psychosis; nausea; vomiting; abnormal choreiform movements; dystonia; on–off phenomenon
Dopamine receptor agonist					
Bromocriptine mesylate	Parlodel	2.5- and 5-mg tablets and capsules	5–100 mg	Activates dopamine receptors	Hypotension; nausea; vomiting; limb edema; toxic psychosis

to 100 mg/1000 mg. The drug is best given in four equally divided doses while awake. Sedation can occasionally limit daytime use. Caution should be advised regarding driving or use of dangerous machinery while medications are being titrated. Side effects can be severe and limiting. Abnormal involuntary movement, confusion, toxic psychosis, nausea, and vomiting can result; they usually resolve on reducing dosage. Combination therapy with bromocriptine is currently being suggested in patients having difficulty being controlled on this form of therapy alone.

Amantadine HCl (Symmetrel®) was first introduced as an oral antiviral (influenza A) agent. Although its mechanism of action is not completely understood, it is felt to increase dopamine release within the CNS. Doses of 100 mg b.i.d. or t.i.d. usually result in clinical improvement within 48 to 72 hr. Although it usually is reserved for early-stage Parkinson's disease, it is also a useful adjunctive medication when the usual dose of levodopa is no longer sufficiently effective or where levodopa dosage had to be reduced because of side effects. Side effects of amantadine include confusion, edema, and livedo reticularis over the limbs.

Tricyclic compounds can be useful agents in early parkinsonism. They not only block reuptake and storage of dopamine but also have anticholinergic properties. Ten milligrams of imipramine or 25 mg amitriptyline, t.i.d. or q.i.d., is the usual dose required for therapeutic effect.

Although anticholinergic agents can be beneficial in reducing symptoms of parkinsonism, they have significant risks and are poorly tolerated by many elderly, making their use less desirable.

Potential Side Effects of Anticholinergic Agents

1. Decreased salivary flow
2. Cardiac arrhythmia
3. Postural hypotension
4. Bowel atony
5. Atonic bladder with potential for overflow incontinence
6. Blurred vision
7. Dementia/confusion/psychosis

Antihistamines such as diphenhydramine HCl (Benadryl®) in doses of 25–200 mg a day can be therapeutically helpful by counteracting the muscarinic effect of acetylcholine in the CNS. Benefit–risk ratio must be carefully considered. Sedation and anticholinergic side effects must be closely watched for.

Bromocriptine (Parlodel®) , a dopamine agonist, is the most recent addition to the pharmacological armamentarium available for the treatment of parkinsonism. Bromocriptine is thought to act by increasing the dopamine content in the CNS by decreasing hypothalamic and striatal dopamine turnover. A starting dose of 1.25 mg b.i.d. with meals is suggested. Dosage may be increased every 14 to 28 days by 2.5 mg per day with meals. Individualized assessment is essential to insure that the lowest dosage producing an optimal therapeutic response is not exceeded. Side effects may include nausea, headache, abnormal involuntary movements, hallucinations, confusion, "on–off" phenomenon, dizziness, hypotension, fatigue, vomiting, abdominal cramps, nasal congestion, constipation, and drowsiness.

Recently, combination therapy (mean dose 7.5–20 mg/day) with levodopa has been suggested as a way to obtain therapeutic responses with lower dosages of both medications. Fewer side effects are to be expected with this form of therapy as well as fewer fluctuations in disability. In general, it is suggested that bromocriptine be started after levodopa has been titrated to 300–400 mg/day. When used alone, bromocriptine is usually required to be given in a dose of 30–50 mg a day; a maximum of 300 mg a day can be used if necessary, and side effects do not limit its use.

IV. PERIPHERAL NEUROPATHIES

The peripheral nerve is that portion of the motor–sensory unit that begins at the junction of the dorsal and ventral roots of the spinal cord and extends to the neuromuscular junction. Problems may result from diseases affecting function (neuropathy), compression (entrapment syndromes), or trauma.

A peripheral neuropathy must be considered in all patients complaining of impaired sensation, i.e., numbness, tingling, and/or burning. Although most cases present symmetrically, an asymmetric presentation, with or without motor involvement, may be noted. Weakness, if also present, is usually distal. Rarely, diabetic myopathy presents with a proximal weakness.

Etiology of Peripheral Neuropathies

1. Drugs and toxins   6. Hereditary
2. Nutritional   7. Tumor
3. Alcoholic   8. Porphyria
4. Diabetic   9. Infectious
5. Amyloidosis   10. Guillain–Barré syndrome

In evaluating a neuropathy, it is important to determine whether the onset has been acute or subacute; the degree of motor versus sensory involvement; whether the symptoms are relapsing; the distribution; whether there is any suggestive family history; and medications used.

Clinical examination can help in making a diagnosis. Selective nerve involvement occurring asymmetrically suggests the diagnosis of a mononeuritis multiplex associated with diabetes mellitus. The femoral or median nerve is usually initially involved, with other nerves being affected as the disease progresses.

Periarteritis nodosa usually affects a single peripheral nerve. Cranial nerve involvement, primarily the oculomotor nerve sparing pupillary fibers, also suggests a diabetes-related neuropathy. Cranial nerve dysfunction involving the oculomotor and lateral rectus nerves has been reported to occur in cases of infectious mononucleosis. Autonomic involvement including cardiac rhythm disturbances, orthostatic hypotension, alteration in diaphoresis, abnormal pupillary reaction, constipation or diarrhea, bladder involvement, and impotence have been associated with diabetes, amyloidosis, and Guillain–Barré syndrome. Associated findings must be carefully looked for. These include alopecia in thallium intoxication; "mees" lines on the fingernails in arsenic poisoning; gingival hyperplasia in phenytoin-induced neuropathy; macroglossia in amyloidosis; and tongue glossitis in pernicious anemia.

Signs of weakness, atrophy, and diminished or absent reflexes suggest a lower motor neuron dysfunction. Nerve conduciion studies are usually required to confirm a suspected diagnosis. Examination of cerebral spinal fluid may be necessary to document Guillain–Barré syndrome. Suggested laboratory screening includes complete blood count, serum electrolytes, BUN, creatinine, B_{12}, folate, glucose, T_4, T_3RU, rheumatoid facfor, antinuclear antibody, anti-DNA antibody, chest X ray, and urinalysis.

In suspected cases, hair and nail analysis can be done to rule out arsenic poisoning and urine sent for heavy metal determinations. Porphyria can be ruled out by measuring δ-aminolevulinic acid in the urine. Other tests that may be helpful include liver function, Schilling test, and nerve biopsy.

Treatment should be directed at the underlying cause of the neuropathy as well as dealing with any associated discomfort. Analgesics are usually effective up to a point; physical activity appears to improve functioning. Although phenytoin has been helpful in certain cases of diabetic neuropathy, no one agent has been proven to be globally effective.

Medications Associated with Neuropathy

1. Nitrofurantoin
2. Dilantin
3. Vincristine
4. Isoniazid
5. Corticosteroids
6. Hydralazine
7. Allopurinol
8. Insulin

(See table on p. 338.)

V. CEREBRAL ISCHEMIA/INFARCTION

Cerebral vascular accidents (CVA, stroke) are the third leading cause of death in the United States. Despite this, mortality from CVA has declined in recent years. Of the 400,000 annual deaths related to CVA, approximately 40% die within 1 month. Two-thirds of those who survive suffer significant impairment; 10% require institutionalization. Although a CVA may occur at any time, 75% occur in those over 65 years of age.

Treatment must be individualized, with the goal being to maintain independent functioning and prevent recurrence.

Characteristics of Selected Neuropathies

Neuropathy	Family history	Acute onset	Subacute onset	Predominantly motor	Mixed sensory motor	Relapsing	Symmetrical	Asymmetric
Guillain–Barre	−	+	−	+	−	−	+/−	+/−
Diabetes mellitus	+/−	+	−	−	+	−	+	+
Periarteritis	−	+	−	−	+	−	−	+
Paraneoplastic syndrome	−	+	+	+	−	−	+	+
Hypothyroidism	+/−	−	+	−	+	−	+	−
Thallium poisoning	−	−	+	−	−	−	+	−
Vitamin B$_{12}$ deficiency	−	−	+	−	−	−	+/−	+/−
Alcoholism	−	−	+	−	+	−	+	−
Heavy metal intoxication	−	−	+	−	+	−	+	−
Drugs	−	+/−	+/−	−	+	−	+	−
Amyloidosis	+/−	−	+	−	+	−	+/−	+/−
Sarcoidosis	−	−	+	−	+	−	−	+
Uremia	−	−	+	−	−	−	+	−
Porphyria	+/−	−	+	+	−	+	−	+

Differential Diagnosis of Stroke

History	Suspect
Trauma	Epidural or subdural hematoma
Subacute onset	Mass lesion
Fever or leukocytes	Brain abscess
Headache or neck discomfort	Subarachnoid hemorrhage
Seizures	Focal seizure
Diabetes/dehydration	Hyperosmolar state
Migraine	Migraine headache

Risk Factors Associated with CVA

1. Hypertension
2. Heart disease
3. Transient ischemic attacks
4. Diabetes mellitus
5. Advanced age

Widespread treatment of hypertension is thought to be the major cause of the reduced mortality from stroke in recent years. Transient ischemic attacks must be evaluated thoroughly. Thirty-five to sixty percent of patients having TIAs will have a completed stroke within 5 years. More than 20% will have a stroke within 1 month, and 50% within 1 year. The chance of having a stroke increases thereafter by approximately 5% per year.

Whereas "nonvalvular" atrial fibrillation has been associated with a 5% increase in the rate of stroke, atrial fibrillation associated with valvular rheumatic heart disease is associated with an 18-fold increase. The post-myocardial-infarction patient also has an increased chance of developing a stroke. Although this risk is still small (2.5%), two-thirds of those who develop a stroke do so within the first few weeks post-MI. Mitral valve prolapse, congestive cardiomyopathy, and coronary disease with or without atrial fibrillation are also associated with an increased risk of developing a stroke.

Characterization of Stroke

1. Completed stroke
2. Stroke in evolution
3. Transient ischemic attack

Persistent focal neurological dysfunction connotes a completed CVA. A transient ischemic attack is diagnosed if the dysfunction remits within 24-hr; a reversible ischemic neurological dysfunction (RIND) is characterized by findings lasting greater than 24-hr. A stroke in evolution is associated with a deterioration of paresis, speech, and sometimes level of consciousness.

Localization of a Stroke

Location	Clinical Findings
Posterior circulation	
Brainstem	
Somewhere in brainstem	Crossed hemiparesis: motor deficit contralateral to the cranial nerve findings
Pons	Pinpoint pupils with ocular bobbing
	Monoparesis (arm) plus dysarthria
Medulla, lateral	Facial numbness; limb ataxia; Horner's ipsilaterally; diminished pain and temperature on contralateral side
Cerebellum	Ataxia, nystagmus (early)
	Conjugate gaze palsy; ipsilateral fifth and peripheral seventh nerve palsies (late)
Anterior circulation	
Cortical	
Left hemisphere	Aphasia
Nondominant hemisphere	Hemiparesis without aphasia but evidence of cortical sensory loss (two-point discrimination, neglect)

(continued)

Localization of a Stroke

Location	Clinical Findings
Frontal cortex	Hemiparesis (leg > arm)
Occipital lobe	Visual field deficit (homonymous hemianopsia) with little or no motor involvement
Subcortical	
Internal capsule	Unilateral motor deficit (face = arm = leg)
Basal ganglia	Dystonic posturing
Thalamus	Sensory loss without motor involvement
Internal capsule or Pons	Motor deficit without sensory loss
	Monoparesis (leg) with ataxia

Transient Ischemic Attack: Localizing Signs and Symptoms

Carotid Vasculature (Anterior Circulation)

1. Amaurosis fugax (visual loss of eyesight on ipsilateral side)
2. Aphasia
3. Transient motor dysfunction in single extremity
4. Transient sensory dysfunction in single extremity

Vertebral Basilar System (Posterior Circulation)

1. Dysarthria
2. Dysphagia
3. Diplopia
4. Vertigo
5. Nausea
6. Vomiting
7. Loss of consciousness
8. Crossed motor or sensory loss

Classification of CVA by Type

1. Large-vessel thrombosis
2. Small-vessel thrombosis (lacunar)
3. Embolism
4. Hemorrhagic

Clinical Features of CNS Involvement

Clinical features	Large-vessel thrombosis	Small-vessel thrombosis (lacunar)	Embolism	Hemorrhage
Onset	Gradual, step-wise, or stuttering	Abrupt or gradual	Sudden	Gradual over minutes or days; sudden
Headache at onset	+ Yes	No	+++ Yes	+++ Yes
Vomiting	++ Yes	+ Yes	+ Yes	+++ Yes
Seizures	+ Rare	+ Rare	++ Yes	++ Yes
Coma	+ Yes	+ Yes	+ Yes	+++ Yes
Preceded by TIA?	++ Yes	No	No	No
Sleep	Asleep	Asleep	Awake	Awake
Associated risk factors	Hypertension Atherosclerosis	Hypertension	Atrial fibrillation	Hypertension

A. Evaluation of the Stroke Patient

The unenhanced CT scan can help identify an acute brain infarction within 24 to 48 hr and a hemorrhage immediately. For atypical presentations or infarctions that are associated with a mass effect (15% of cases), a dye-enhanced scan may be required to exclude a tumor. Unfortunately, for technical reasons, the CT scan has difficulty imaging the posterior fossa (brainstem and cerebellum).

Magnetic resonance imaging (MRI), although not yet widely available, appears to be far superior in identifying lesions within the posterior fossa. It is, however, unable to distinguish acute infarction from hemorrhage; although it is highly sensitive, it is nonspecific.

The use of MRI with contrast agents should improve its diagnostic usefulness.

Ophthalmodynomometry (ODM), continuous-wave Doppler ultrasonography, and facial thermography are useful screening procedures to evaluate for pathology at the carotid bifurcation. The Doppler has an 80% correlation to carotid angiograms and has no associated morbidity.

Invasive diagnostic procedures should be restricted to patients who are considered to be surgical candidates. Age, in itself, is not a contraindication to surgery.

Digital subtraction angiography (DSA) using either intravenous or intraarterial studies can be used to delineate the carotid bifurcations. One limitation, however, is the potential to miss lesions less than 50% occlusive. In addition, the intravenous approach requires relatively large volumes of dye, increasing the risk of congestive heart failure, angina, or renal insufficiency. Although the intraarterial approach requires only a small volume of dye, it is associated with risks similar to those of direct cerebral arteriography, i.e., embolism and hemorrhage. Cerebral arteriography is the preferred technique in most centers and can be used to study anterior, posterior, and aortic arch pathology.

Examination of the cerebral spinal fluid is indicated in cases in which infection or subarachnoid hemorrhage is suspected in the absence of focal neurological signs. A CT scan is suggested prior to doing the lumbar puncture to rule out mass lesions.

Laboratory tests should include a CBC, prothrombin time, partial thromboplastin time, VDRL, blood sugar, lipid profile, electrocardiogram, and chest X ray.

Controversy exists regarding when to admit the patient with cerebral vascular disease to the hospital. Most agree that completed CVA and acute TIA warrant admission to the hospital for rehabilitation and further evaluation. Although patients who had a TIA more than 1 week ago usually are evaluated as outpatients, it is important to remember that 20% will progress to a complete stroke within 1 month.

B. Treatment of Stroke

Rapid stabilization and evaluation
Immobility must be dealt with early; begin a rehabilitation program promptly
Prevention

1. Stabilization

1. Document clinical and functional status.
2. Insure a patent airway. Vomiting, lethargy, and inability to eliminate secretions can become life-threatening. A nasogastric tube or intubation may be required in select cases. Pneumonia is the leading cause of death in stroke patients and must be actively prevented.
3. Avoid solutions with high concentrations of free water and glucose. These have been associated with cerebral edema.
4. Avoid treating mild to moderately elevated blood pressures in the setting of an acute CVA. Blood pressure usually declines within 6–12 hr post-CVA. Rapid blood pressure drop or hypotension can significantly worsen cerebral perfusion. In general, blood pressure should be treated only if greater than 160–170/95–100 in the acute phase of the CVA. If blood pressure lowering is required, a slow and carefully monitored course is advised, preferably with arterial line monitoring. Nitroprusside, i.v. drip, can be closely titrated to individual needs.
5. Consider steroids. The role of dexamethasone in reducing cerebral edema remains controversial.

2. Immobility

1. Early and aggressive rehabilitation may reduce contractures, pressure sores, and thromboembolic disease. Incontinence, pneumonia, and urinary tract infection may also be prevented.
2. Change position every 2–3 hr.
3. Passive range of motion should be done three to four times a day.
4. Continue to provide emotional support and encouragement.

3. Prevention

1. Review all risk factors and treat when appropriate.
2. Hypertension should be controlled; avoid rapid changes during the 2 to 3-week peristroke period.
3. Any TIAs in the anterior and posterior circulation should be treated with antiplatelet agents. Aspirin, 300 mg daily, and in some cases persantine, 75 mg t.i.d., have been suggested.

4. Pursue noninvasive tests to determine whether a surgical approach is warranted.
5. Reduce hematocrits greater than 50% by treating underlying cause.
6. Reduce blood sugar to 120–140 mg/dl fasting and 160–200 mg/dl postprandial.
7. Discontinue cigarette smoking.

Indications for Anticoagulation in Cerebral Vascular Disease

1. Completed stroke secondary to cerebral emboli of cardiac origin
2. Stroke in progress (controversial)
3. Flurry of TIAs (controversial)

Contraindications for Anticoagulation

1. Liver failure
2. Renal failure
3. Advanced age (controversial)
4. Frequent falls
5. Bleeding diathesis
6. Hypertension (uncontrolled)

4. Surgical Considerations

Surgical treatment consists of a carotid endarterectomy. Recent studies indicate that an external-to-internal carotid bypass is ineffective in preventing strokes. Patients with TIAs in the carotid distribution with stenosis of greater than 70%, however, should be considered possible surgical candidates. Patients with completed strokes in the anterior distribution whose residual deficit is minimal both clinically and diagnostically (CT scan) are surgical candidates if significant stenosis is noted in the affected carotid.

In the elderly, risk factors must be carefully weighed against possible benefits and expected lifespan at time of surgery. Emergency surgery may be required for removal of blood clots in the posterior fossa that show evidence of expansion regardless of patient's age. Surgical intervention is not warranted in patients who have had a completed stroke less than 5 weeks ago, a large stroke, or those having TIAs in the posterior circulation.

VI. DEMENTIA

A. Introduction

Dementia is a syndrome characterized by impaired intellectual function. This is accompanied by memory loss and altered personality of sufficient severity to impair social or occupational functioning. Although the term usually implies a progressive and generally irreversible deterioration of mental functioning, at least 25% of the cases have a treatable organic basis. Dementia is the number-one cause of nursing home placement. It not only exhausts the physical, economic, and mental resources of the person affected but also those of the family unit. It is important to recognize that dementia is not a part of normal aging. Only 2.3% of persons aged 65 to 69 are diagnosed as having dementia; this number rises to 5.5% of those 75 to 79 and over 20% of those over 80.

B. Senile Dementia of the Alzheimer's Type

Over 50% of the cases of dementia can be characterized as senile dementia of the Alzheimer's type (SDAT). The pathological changes found in the brain are indistinguishable from those of Alzheimer's disease, a genetically linked condition resulting in dementia usually beginning in the sixth decade of life. Pathological change are characterized by an excessive number of neurofibillary tangles and senile plaques within the brain matter. Because it is impossible to make the diagnosis without actually obtaining brain tissue for histology, it is imperative before a person is labeled as having SDAT that other treatable and reversible causes of dementia be excluded.

Recent studies have shown altered biochemical parameters in brains of patients with SDAT. Diminished activity of the enzyme choline acetyltransferase, with a resulting decrease in acetylcholine (a potent neurotransmitter thought to be intimately related to memory), is just one of the many notable changes that have been detected.

Typically, the exact onset of memory loss and/or behavioral change is hard to pinpoint. Early on, loss of memory for recent events and in the ability to concentrate are noted. Confabulation often becomes a way of compensating for worsening memory. As the dementia becomes more advanced, thought processes appear diffuse, and the patient may become disoriented. Significant changes are noted in social behavior, with increased aggression, assaultiveness, wandering, and apathy common. Patients with incipient SDAT may be depressed over the realization that their mental functioning is deteriorating, whereas those with advanced dementia exhibit little insight or fear of their altered state.

C. Multi-infarct Dementia

This form of dementia (MID) is responsible for approximately 25% of cases of dementia. Multiple infarcts secondary to thrombosis, hemorrhage, or emboli are responsible for permanent brain tissue damage. A destruction of 50 ml of brain tissue appears to be a critical volume beyond which clinical manifestations of dementia are detected. Unlike SDAT, where onset is insidious and the disease progresses slowly over time, MID often progresses in a stuttering or stepwise fashion. Day-to-day variation in mental functioning may also be seen.

Distinguishing Characteristics

	MID	SDAT
Onset	Rapid	Slow
Course	Stepwise	Progressive
Etiology	Multiple lacunar infarcts	Cortical degeneration
Findings	Fluctuating mental status	Global deterioration
	Possible focal neurological signs	No physical signs
	History of hypertension and/or cardiovascular illness	No significant history
	Men > women	Women > men
Treatment	Correct underlying disorder for prevention of further deterioration	Symptomatic

Multi-infarct dementia occurs more frequently in men, and a history of hypertension is common. Other than for occasional neurological deficits, the signs and symptoms of MID are almost indistinguishable from those found in SDAT. It is important to remember that the two pathological processes can coexist.

D. Potentially Reversible Dementia

Approximately 25% of the cases of dementia are clearly preventable. Early detection and prompt treatment determine the degree and extent of reversibility. Although causes for this type of dementia are multiple, general categories include:

Drug toxicity
Emotional disorders
Metabolic and endocrine abnormalities
Eye and ear problems
Nutritional disorders
Tumor and trauma
Infections
Atherosclerotic complications

Reversible Causes of Dementia

Medication problems
Endocrinopathies
   (hyper- and hypothyroidism; hyper- and hypo-
   glycemia; hypercalcemia)
Salt and water imbalance (hypo- and hyperatremia)
Mass lesions (neoplasm, subdural hematoma)
Vitamin B_{12} deficiency
Depression
Relocation
Disease of the heart and lungs
Renal and hepatic insufficiency
Normal pressure hydrocephalus
Infections of the CNS
Communication problem
   (decreased hearing, low vision, aphasia, etc.)
Toxins (alcohol, heavy metals, carbon monoxide, etc.)

Perhaps one of the most important and often forgotten differential diagnoses of dementia is depression. The following summarizes some of the classic factors that can help distinguish between these two entities. Overlap between these syndromes is frequently seen, and the following is only a guideline in helping to differentiate between the two.

Depression versus True Dementia

	Depression	Dementia
Onset	Rapid; exact onset can often be dated	Insidious and ill-defined
Behavior	Stable; depression, apathy, and withdrawal common	Labile; fluctuates between normal and withdrawn and apathetic
Mental competence	Usually unaffected; however, may appear demented at times; complains of memory problems	Consistently impaired, tries to hide cognitive impairment
Somatic	Vegetative signs common: anxiety, insomnia, eating disturbances, minor physical complaints	Occasional sleep disturbances
Self-image	Poor	Normal
Duration	Usually self-limited; reversible with therapy	Chronic; slowly progressive

E. Evaluation

The burden is on the clinician to rule out all readily treatable forms of "dementia" prior to labeling the elderly person as having an irreversible condition. A complete evaluation should be done as early as possible. Delay in diagnosing dementia with a reversible cause, such as hypothyroidism, may result in a worsening of the condition. In addition, failure to treat the disease early in its course may also lessen the chances for complete recovery because of additional brain damage from long-term lack of thyroid hormone.

Every evaluation should include a thorough history. Family and/or care givers should be interviewed with and without the patient present to obtain as much and as accurate information as possible. Information should be sought regarding the onset of symptoms, any associated complaints, and/or other illnesses present. A complete history of medication usage is essential. Other important historical factors are listed. As mentioned above, particular attention should be given to rule out depression.

Important Historical Factors

Duration of signs and symptoms
Pattern of progression of illness
Medication usage
Alcohol intake
Family history
Any witnessed aberrant behavior
Other medical or neurological problems
Ability to perform activities of daily living
Risk factors, i.e., falls, wandering, etc.
Stress on care givers

A comprehensive physical examination should be performed. Mental status traditionally is determined by a series of standardized tests. The FACT test is an example of one such test. The patient is asked to list ten animals, ten colors, ten fruits, and ten towns. In the absence of a language or neurological impairment, fewer than 15 to 20 correct answers out of an optimal score of 40 suggests a seriously demented state. Dementia should be suspected if the score is less than 30. Laboratory workup is aimed at assessing those treatable conditions that may result in dementia. Hemoglobin, folate and B_{12} levels, thyroid function tests, blood sugar, serology for syphilis, serum calcium, phosphorus, electrolytes, liver function tests, and creatinine are recommended. Most authorities agree that an evaluation for dementia is not complete without a computerized axial tomography scan (CT scan) of the brain; an electroen-

cephalogram, and possibly a lumbar puncture. The finding of large
ventricles and wide sulci on CT scan does not necessarily indicate
dementia; test results must not be misinterpreted. Specific drug
levels should be ordered as clinically indicated.

Laboratory Screen for Dementia

Complete blood count
Serum chemistries (including electrlytes, renal indices,
    liver function tests, calcium, phosphorus, total protein,
    albumin)
Thyroid function tests
Serum B$_{12}$ and folate levels
Urinalysis
VDRL
Chest X ray
ECG
Computed tomography of the brain with contrast

F. Nonpharmacological Management

Despite the magnitude of the problem, medical science has
been unable to find a cure for the majority of cases of dementia. It
therefore becomes increasingly important to recognize treatable
causes of this illness.

Most patients with dementia are cared for in their own homes,
often placing a great deal of stress on the care givers. Family stress
often limits the ability of the demented person to remain at home.
Whereas most care givers are willing to tolerate problems of urinary
incontinence and falling, few are able to deal with problems of
daytime wandering and interference with sleep. Although stress
seems to be more of a factor in those families where interpersonal
relationships were poor to begin with, there comes a tolerance point
beyond which families can no longer meet the needs of the
demented person. In order to lessen this burden of responsibility
and maintain a home-based environment, it is essential that all
avenues of assistance be explored, including meals-on-wheels,
visiting nurse, respite care, day care, etc. In certain states, formalized "community options" programs have been initiated to
facilitate and coordinate services.

Depending on the individual needs of the patient with advanced dementia, admission to either a nursing home, hospital, or psychiatric facility may be required. Unfortunately, this decision often leads to family guilt and at times compromised care. Admission is often difficult to arrange because of the lack of insight that these patients have regarding their problems and needs. Skilled health professionals should be consulted, as they can offer invaluable assistance to patients and their families during these difficult times.

Once admitted to the hospital or nursing home setting, a high degree of care is usually required. Deliberate efforts should be made to provide a degree of orientation through verbal or visual cues (e.g., calendar, clock, orientation board). In addition, color schemes can be useful aids in helping the patient identify his/her own room. Independence should be maintained as much as possible, and the possession of personal belongings encouraged. Formalized activities help prevent apathy, isolation, depression, and physical inactivity and thus should be encouraged.

Often overlooked is the role of guardianship. Once a patient is no longer capable of understanding what is occurring in his surroundings and making formal decisions, a guardian needs to be appointed. This should be someone familiar with the patient's premorbid wishes and desires and who is judged to have the patient's welfare at heart. Once a guardian is identified, consultation can be sought regarding what evaluation, therapies, and placements are most appropriate for a given patient.

G. Drug Therapy

If an underlying and potentially reversible cause of dementia is diagnosed, specific therapy to correct the problem is indicated. In cases of nonreversible dementia, a multitude of medications are available to treat many of the associated behavioral problems. The prescribing clinician should become knowledgeable of specific drug side effects and how age may modify drug pharmacokinetics.

Major tranquilizers may provide an "orienting" effect as well as improving aberrant behavior. Commonly used drugs in this category include haloperidol (Haldol®), thioridazine (Mellaril®), and thiothixene (Navane®). Small dosages are often very effective in this patient population; drug benefit versus drug side effect must be carefully weighed.

Antidepressant medications can be a useful clinical adjunct in treating patients with pseudodementia resulting from depression and in controlling depressive symptoms in general. Once again, side effects must be carefully monitored.

Currently available pharmacological therapies, for the most part, provide symptomatic relief only and do not reverse or slow the progression of dementia. Drugs (vasodilators) used in an attempt to increase cerebral blood supply have not proven effective in slowing progression of or reversing dementia or in improving memory.

In recent years, intense investigations have begun to provide some signs of therapeutic hope. Based on the "cholinergic hypothesis" of memory and the finding of reduced acetylcholine content in brains of patients with Alzheimer's disease, choline and lecithin have been administered orally. Although some have reported positive results, for the most part studies have not been as rewarding as was hoped. Questions have arisen as to dosage and the ability of these agents to be incorporated into functional brain tissue. Another approach to this treatment is by increasing acetylcholine levels in the CNS by inhibiting its metabolism. Translating preliminary research data gathered from a small number of subjects into widespread effective therapy, however, remains a challenge.

Dehydroergotoxine mesylate (Hydergine®), an ergot alkaloid, has been reported in numerous case studies worldwide to have therapeutic efficacy if given early in the course of dementia. Improvement in function, reduced associated symptomatology, and slowing of the disease progression have been reported. Although this agent has been shown to have numerous potential actions on the central nervous system, dehydroergotoxine is now considered to benefit brain metabolism indirectly through its influence on cerebral enzymes. Since this agent has minimal side effects, many recommend a trial of this drug for patients with early signs of dementia not caused by otherwise treatable illness. In the United States, 1 mg t.i.d. is the accepted dosage. Preliminary data from other countries suggest the need for a higher dose to obtain maximal effects. In all cases, a minimal trial period of 3 to 6 months should be given. Because of the nature of the disease, failure of the disease to progress while on medication may in itself signify therapeutic success. Although a limiting factor remains this drug's relatively expensive cost not covered by state and federal programs, one must consider the potential savings that would result from a successful trial of this drug.

In patients with multiinfarct dementia and coincident hypertension, hypertension should be treated in an attempt to arrest the neurological damage and dementing process.

At present, many drugs, including piracetam, vincamine, and "active lipid," are undergoing clinical trials.

H. Research

Several experimental models presently exist for the study of SDAT and its pathogenesis.

Drugs that block acetylcholine receptors, such as scopolamine, induce memory deficits in healthy young subjects similar to those observed in aged, nondemented individuals. These deficits can be reversed by administering physostigmine, a cholinergic agonist capable of potentiating the synaptic effects of released acetylcholine. Increased levels of aluminum have been reported in brains of patients with Alzheimer's disease. In further support of a causal relationship, aluminum salts administered to cats result in pathological changes resembling those found in Alzheimer's disease. Additional research is clearly needed.

Experimental Models of Dementia

1. Aged animals
2. Aluminum toxicity
3. Hypoxia
4. Anticholinergic drugs
5. Destruction of dopaminergic pathways in CNS
6. Viral infections

I. Summary

Senile dementia of the Alzheimer's type is responsible for the majority of cases of dementia. At least 25% of dementia, however, may have a preventable and reversible cause if found early in the course. Benign senescent forgetfulness must not be confused with dementia. The loss of a certain amount of short-term memory is a normal aging process; this loss, however, does not interfere with one's ability to conduct activities of daily living.

Dementia is the number-one cause of patients entering nursing homes today. Although fewer than 3% of people in the seventh decade of life have dementia, approximately one in five of those over 80 years of age are affected. Women are affected twice as commonly as men. Once SDAT is diagnosed, one's projected life expectancy is reduced by one-third to one-half. Although SDAT has a higher prevalence in certain families, failure to develop the disease by the mid-60s appears to reduce the person's risk to that of the person without a positive family history. A thorough evaluation is essential to rule out treatable causes of dementia prior to labeling someone as having either SDAT or MID.

Although therapeutic measures are limited, family support, behavior modification, and drug treatments should be explored. Additional research should improve our understanding of this devastating illness and offer new avenues of prevention and treatment.

VII. DELIRIUM

Delirium can be defined as a problem of both cognition and level of consciousness. It is seen frequently in the hospitalized elderly and can be a life-threatening complication of any acute illness. Recognition of the condition is essential in order to assure an early diagnosis and treatment. Delirium is associated with a very poor prognosis primarily because of the severity of accompanying medical illness.

A. Diagnostic Criteria for Delirium

1. Clouding of consciousness (reduced clarity of awareness of the environment) with reduced capacity to shift, focus, and sustain attention to environmental stimuli.
2. At least two of the following:
   a. Perceptual disturbance: misinterpretations, illusions, or hallucinations.
   b. Speech that is at times incoherent.
   c. Disturbance of sleep–wake cycle, with insomnia or daytime drowsiness.
   d. Increased or decreased impairment (if testable).
3. Disorientation and memory impairment (if testable).
4. Clinical features that develop over a short period of time

(usually hours to days) and tend to fluctuate over the course of a day.

5. Evidence, from the history, physical examination, or laboratory tests, of a specific organic factor judged to be etiologically related to the disturbance.

The major condition that mimics delirium is dementia, with many of the same diagnostic features. The major differentiating feature, however, is the impaired level of consciousness associated with delirium. Clinically, it is often difficult to distinguish between the two. The necessity of an accurate diagnosis cannot be overemphasized.

B. Conditions Leading to Delirium in the Elderly

1. Medications: numerous medications can lead to delirium. Common examples include antiarrhythmics, antihypertensives, cardiac glycosides, antiparkinson agents, tricyclic antidepressants, neuroleptics, benzodiazepines, steroids, and many others.
2. Metabolic: hypoxia, hypercarbia, thyroid dysfunction, uremia, hepatic encephalopathy, congestive heart failure, hypotension, electrolyte disturbance, acid–base disturbance, hypovitaminosis, hyperglycemia, hypoglycemia, and many others.
3. Sensory and/or sleep deprivation: seen in "ICU psychosis," "sundown syndrome."
4. Tumor: any form of malignancy may result in mental status changes suggestive of delirium.
5. Trauma, particularly closed-head trauma resulting in subdural hematoma, concussion.
6. Infection: any infection may cause mental status change; meningitis and encephalitis are most often implicated in delirium.
7. Exogenous toxins, particularly alcohol, carbon monoxide, heavy metals.

The management of the delirious patient requires identification of the underlying etiology, correction of any potentially reversible causes, physiological support, and control of any agitation that may result from the clouding of consciousness.

C. Management of Delirium

1. Treat any underlying organic etiology.
2. Physiological support of fluid and nutritional balance.
3. Environmental support including constant attention from family and nursing staff; orientation and reassurance; a brightly colored environment with appropriate clocks, radios, and night light.
4. Restraint: physical and/or chemical restraints may be required if agitation places the patient at risk of injury.
5. Low-dose psychotropic medication may be required: 0.5–2 mg of haloperidol p.o. or i.m., one to two times per day, or its equivalent may be helpful; careful titration and observation are essential to avoid undesirable side effects.

VIII. HYPOTHERMIA IN THE ELDERLY

Hypothermia is defined as a core body temperature less than 35°C. The elderly are more susceptible to this problem than the younger population. It is often a life-threatening emergency and requires careful treatment. If the body's core temperature is below 30°C, the mortality is as high as 70%. Between 30 and 33°C, the mortality remains high at 33%.

A. Factors That Predispose the Elderly to Hypothermia

1. Environment. Wet winter conditions as well as subfreezing temperatures promote heat loss.
2. Drugs. Major tranquilizers affect the hypothalamic centers responsible for heat dissipation and shivering. Alcohol promotes peripheral heat loss. Minor and major tranquilizers may cause confusion and allow the patient to remain in the out-of-doors or in hypothermic indoor environments longer than safely acceptable.
3. CNS disease. Stroke, Parkinson's disease, and dementia limit mobility and decision making for environmental changes.
4. Autonomic nervous dysfunction. Age-prevalent illness

and/or medication use may lower sympathetic response to stress.
5. Habits. The elderly are often bundled up in the absence of adequate home heating and may be reluctant to seek warmer shelter or accept warmer shelter from outsiders.
6. Intercurrent/concurrent illness. Pneumonia and/or congestive heart failure may precede or coincide with clinical hypothermia.
7. Miscellaneous: myxedema and Addison's disease.

B. Physiology of Hypothermia in the Elderly

1. Skin receptors for cold outnumber heat receptors 10 : 1.
2. The anterior hypothalamus is the heat loss center, causing generalized vasodilation.
3. The posterior hypothalamus is the heat conservation center, causing vasoconstriction of the skin, vasodilation of muscle, shivering, and increased metabolic rate.
4. Shivering raises metabolic rate four to five times normal by raising muscle temperature. Unfortunately, muscle mass is lower in older persons, resulting in less efficient shivering thermogenesis.
5. Under normal homeothermic conditions, muscle temperature is 1°C higher than skin temperature. Visceral–core temperature is 0.5–1°C higher than muscle temperature. Brain temperature is 1°C cooler than core temperature.
6. With surface cooling, the brain is 1–2°C warmer than the core temperature.
7. With bloodstream cooling, the brain is 4–12°C warmer than the core temperature. Cortical blood vessel constriction, the presence of the cranial vault and venous sinuses, and preferential perfusion compared to other organs allows brain preservation of heat.

The clinical presentation of hypothermia is often nonspecific and may remain unrecognized unless a core temperature is obtained. Often a mild alteration in mental status or a sinus bradycardia may be the only presenting sign. In a mild case of hypothermia (32°C), when the body is actively trying to conserve heat, the skin may be cold to the touch, particularly in the abdomen.
In moderate cases of hypothermia (28–32°C), the body now

begins to show signs of a slower metabolic process resulting in a bradycardia, hypotension, hypoxia, lethargy, and mental confusion. In the most severe cases of hypothermia, when the core temperature drops below 28°C, coma with dilated pupils, hyporeflexia, and no spontaneous movement is recorded. The patient may appear to be clinically dead; resuscitation, however, is always recommended in these cases.

Clinical Presentation by System of Hypothermia in the Elderly

1. Central nervous system
   a. 34–33°C Sensorium decreased; deep tendon reflexes increased; pupillary responses increased.
   b. 32°C EEG voltage decreased, parietal and occipital
   c. 30°C Sensorium further depressed; CSF pressure decreased; deep tendon reflexes and pupillary responses decreased; enhanced evoked potentials; EEG frontal lobe δ and θ wave activity increased; β and α activity depressed
   d. 28°C Evoked potentials normal; aortic and carotid baroreceptor response decreased
   e. 25°C θ activity disappears; evoked potentials decreased; deep tendon reflexes and pupillary responses absent; lower limit of carbon dioxide chemoreceptor response
   f. 21–20°C EEG spontaneous activity disappears; chemoreceptor response to hypoxia ceases; baroreceptor responses abolished
2. Metabolic–endocrine
   Shivering
   a. 36–30°C Excitatory
   b. 27°C Minimally depressed
   c. 25°C Moderately depressed
   ACTH
   a. 28°C Endogenous secretion ceases
   b. 23–20°C Adrenal response to exogenous ACTH ceases

(continued)

Clinical Presentation by System of Hypothermia in the Elderly (Continued)

TSH
a. 31°C   Endogenous production of thyroid hormone depressed; maximum TSH secretion noted
3. Gastrointestinal
Pancreas
a. 31–29°C Insulin decreased; minimal effectiveness
b. 28°C   Insulin secretion ceases; exocrine flow markedly depressed but enzymes normal
Gastrointestinal motility
a. 34°C   Motility decreased
b. 30°C   Motility ceases
4. Respiratory
a. 34°C   Maximum response of respiratory centers to carbon dioxide
b. 32°C   No alteration in pH or in blood gases
c. 30°C   Respiratory quotient decreased from 0.82 to 0.65; decreased oxygen uptake causes decreased carbon dioxide production; decrease in carbon dioxide production is greater than decrease in oxygen uptake
d. 30–28°C Acidosis begins to appear
e. 25°C   Mandatory artificial respiration; anatomic dead space increases 50%, and physiological dead space increases 28% because of cold bronchial dilatation; alveolar dead space is unchanged; rate and minute ventilation decrease less than the decrement in tidal volume
5. Cardiovascular
a. 34–32°C Arterial blood pressure rises
b. 32°C   Decline in arterial blood pressure; cardiac output 75% of normal
c. 30°C   Cardiac output 55% of normal; peripheral resistance increases; minimal standard dose of digitalis is 40% of normal
d. 28°C   Cardiac output 54% of normal; cardiac muscle shortening at maximum efficiency; first-degree heart block; capillary blood flow impeded; plasma volume decreased; impaired baroreceptors

---

(continued)

Clinical Presentation by System of Hypothermia in the Elderly (*Continued*)

    e. 25°C   Cardiac output 30% of normal; depressed cardiac muscle shortening; atrial fibillation as potassium moves into muscle cells; hypotension; peripheral venous pressure normalizes as a result of peripheral sequestration and lymphatic dysfunction; pressor effect of norepinephrine (α receptors) markedly decreased; impaired vasomotor and venomotor reflex mechanisms; minimal standard dose of digitalis 40% of normal

6. Renal
    a. 33°C   Urine flow increased twice normal
    b. 32–30°C Glomerular filtration rate 75%; renal plasma flow 48%
    c. 25°C   Urine flow increased twice normal
7. Hematology
    a. 32–30°C No hematologic changes
    b. 28°C   Liver and spleen sequestration causing lymphopenia and thrombocytopenia; intravascular coagulation, especially in the lungs

C. Treatment

Treatment is largely based on early diagnosis and support of the body functions.

1. Do not load heavy blankets on the patient, since heavy blankets trap cold air
2. Do not use hot water bottles or an electric blanket; peripheral warming causes heat loss and increased cardiac demand
3. Do not encourage movement; increased cardiac demand and rhabdomyolysis can occur
4. Do not give alcohol because of alcohol's peripheral vasodilating action

(*continued*)

5. Do not rub or massage injured peripheral tissues
6. Do not do stomach, colonic, or bladder irrigation with heated fluids
7. If core temperature is greater than or equal to 33°C:
    a. Passive peripheral rewarming with warmed dry blankets to prevent shivering
    b. Heated intravenous fluids at 45°C (113°F) contribute 17 kcal or +⅓°C per liter
    c. Heated humid oxygen at 20 liters/min (unless carbon dioxide retainer) contributes 30–40 kcal/hr
8. If core temperature is less than 33°C and there is cardiovascular instability:
    a. Passive heated blankets raise peripheral temperature 1–2°C/hr and prevent shivering
    b. Peritoneal dialysis with 45°C potassium-free dialysate, 2 liters every 10 min can be used to raise core temperature to 35°C
9. Careful hydration: increased intracellular volume and peripheral sequestration are seen in hypothermia; increased extracellular volume occurs with rewarming
10. Cardiovascular: atrial and ventricular arrhythmias are temperature dependent; minimal standard dose of digoxin must be reduced to prevent toxicity on rewarming
    a. Bradycardia should be treated cautiously with isoproterenol
    b. Ventricular fibrillation should be actively treated
    c. Correct acidosis cautiously with sodium bicardonate only if pH is less than or equal to 7.1
11. Respiratory depression
    a. Increased Pco_2 helps to release oxygen from hemoglobin
    b. Intubation and hyperventilation will improve alveolar and arterial Po_2 but lower alveolar and arterial Pco_2; this may not increase tissue oxygenation until temperature is normalized
    c. Since arterial blood gases are routinely done at 37°C, artifactually lowering pH and raising Pco_2 and Po_2 in the laboratory, correction for actual temperature is required
12. Hyper/hypoglycemia: dextrose should be given in case of hypoglycemia; insulin probably is ineffective for hyperglycemia until temperature is greater than 30–31°C

(continued)

13. Prophylactic antibiotics: in cases of severe hypothermia and coma (after all appropriate cultures have been taken), broad-spectrum antibiotics are indicated

IX. HYPERTHERMIA IN THE ELDERLY

Heat-related illness (hyperthermia) is a major cause of morbidity and mortality in the elderly, with noticeable effects during extended periods of high ambient temperature and humidity. Retirement from work, abundant leisure time, and physical work limitations do not protect the elderly from heat stress. During heat waves, 75% of those elderly hospitalized (or found dead) have heat-related symptoms. During less heat-intense summer periods, heat may contribute to sepsis, myocardial infarction, and cerebrovascular accidents, especially in diabetic patients.

Definition of Hyperthermia

1. Core body temperature greater than 41.1°C (106°F).
2. Core body temperature greater than 40.6°C (105°F) accompanied by altered mental state or anhydrosis
3. Patient dead with liver probe temperature greater than or equal to 41.1°C (106°F)

Stages of Hyperthermia in the Elderly

Stage I: heat edema or heat syncope
    Salt and water deficits
    Hypokalemia secondary to sweating
    Muscular and cutaneous vasodilatation and venous stasis
    Core temperature normal or elevated
Stage II: heat cramps
    Increasing salt deficit
Stage III: heat exhaustion
    Increasing salt deficit
    Nausea and vomiting
    No thirst, temperature normal, sweating present

(continued)

Stages of Hyperthermia in the Elderly (*Continued*)

Hypotension
Decreased or normal urine output
Increasing water deficit
Weakness
CNS excitation or confusion
Blunted thirst response
Temperature increased
Decreased sweating
Oliguria
Stage IV: heat stroke
Increasing water deficit
Increasing salt deficit
CNS dysfunction
Disorientation, confusion, coma, seizures
Temperature elevation (see definitions)
Anhydrosis (but may be variable)
Skin hot, flushed, dry, and cyanotic because of peripheral vasoconstriction
Respiratory alkalosis or mixed metabolic acidosis/respiratory alkalosis
Hepatic necrosis
Mesenteric necrosis
Renal failure uncommon in the elderly
Rhabdomyolysis uncommon in the elderly
Disseminated intravascular coagulation uncommon in the elderly

The elderly, because of physiological changes, are more susceptible to the effects of increased heat and humidity. In addition, problems may result because of a difficulty in identifying the signs and symptoms of dehydration and the decreased thirst mechanism often present in elderly persons.

Physiological Response to Increased Temperature in the Elderly

1. Anterior hypothalamus senses increase in temperature and initiates sweating response; sweating may rise as high as 1500 cc/hr in extreme conditions

(*continued*)

Physiological Response to Increased Temperature in the Elderly (*Continued*)

2. Core temperature rises to dissipate increased heat in peripheral tissues
3. Rapid dehydration occurs as increased sweating occurs
4. Core temperature rises with increasing dehydration
5. Sweating response falls by 30–60% after 3–5 hr in temperatures above 40°C and a relative humidity above 60%
6. Thirst mechanisms (often impaired in elderly) activate after loss of 2% of the body weight via sweating response; oliguria occurs after a loss of 6% of the body weight; coma and impending death occur when dehydration exceeds 15% loss of body weight

Treatment of hyperthermia is largely supportive; maintenance of physiological functions and rapid reduction of core temperature is essential. Unlike hypothermia, where a slow warming (to prevent ventricular arrhythmias) is indicated, a prolonged elevated core temperature results in permanent damage.

Treatment of Hyperthermia

1. CNS. If core temperature is greater than 42°C (107°F), patients' airways should be protected with endotracheal tube; peritoneal lavage with cold potassium-free dialysate (2 liters/10–15 min) until core temperature returns to 38°C; seizure control may be maintained with diazepam as necessary; rehydration with intravenous fluids as necessary
2. Elevated temperature. Do not apply ice directly to skin; this will cause peripheral vasoconstriction and will not dissipate high core temperature; apply ice packs to axilla, posterior neck, and inguinal areas; place a cooling blanket under the patient
3. Oliguria. Urinary output must be maintained, as renal failure is a life-threatening complication; hydration status must be maintained, and the judicious use of osmotic diuretics should be considered
4. Metabolic. Acidosis may occur with tissue necrosis, requiring correction with sodium bicarbonate as neces-

(*continued*)

Treatment of Hyperthermia (*Continued*)

sary; hypokalemia from dehydration should be corrected if urinary output is maintained; hyperglycemia will occur secondary to dehydration and will usually respond to rehydration; insulin is rarely necessary to control hyperglycemia and may aggravate a hypokalemic state

5. Infections. Patients with hyperthermia have a high incidence of infections because of necrosis of tissue and often aspiration pneumonia secondary to obtundation; early recognition and treatment with broad-spectrum antibiotics as necessary is indicated

---

Prognosis of hyperthermia is related to both the peak core temperature attained and the duration of the elevated core temperature.

Poor Prognostic Factors in Hyperthermia

1. Persistently elevated core temperatures despite cooling
2. Altered mental status after temperature declines
3. Markedly elevated liver function tests (SGOT > 1000)
4. Rhabdomyolysis

SUGGESTED READING

Blass J. P.: Alzheimers disease. *DM* 31(4):1–69, 1985.

Calne B.: Current views on Parkinson's disease. *Can J Neural Sci* 10:11–15, 1983.

Cummings J. L.: *Acute Confusional States in Clinical Neuropsychiatry.* Orlando, Grune and Stratton, 1985.

Easton J. D., Sherman D. G.: Management of cerebral embolism of cardiac origin. *Stroke* 11:433–442, 1980.

Hornykiewicz O.: Parkinson's disease, in Crow T. J. (ed): *Disorders of Neurohumoural Transmission.* New York, Academic Press, 1982, pp 121–143.

Kistler J. P., Ropper A. H., Herus R. C.: Therapy of ischemic cerebral vascular disease due to antherothrombosis. *N Engl J Med* 311:27–34, 1984.

Komrad M. S., Coffey E., Coffey K. S., et al: Myocardial infarction and stroke. *Neurology* 34:1403–1407, 1984.

Larson E. B., Reifler B. V., Featherstone H. J.: Dementia in elderly outpatients: A prospective study. *Ann Intern Med* 100:417–423, 1984.

Lipowski Z. J.: Transient cognitive disorders (delirium, acute confusional states) in the elderly. *Am J Psychiatry* 140:1426–1436, 1983.

Pfeiffer R. F., Wilken K., Glaeske C., et al: Low-dose bromocriptine therapy in Parkinson's disease. *Arch Neurol* 42:586–588, 1985.

Terry R. D., Katzman R.: Senile dementia of the alzheimer type. *Ann Neurol* 14(5):497–506, 1983.

Zwiebel W. J., Smother C. M., Austin C. W., et al: Comparison of ultrasound and IV-DSA for carotid evaluation. *Stroke* 116:633–642, 1985.

14

Preventive Health Care for the Elderly

I. INTRODUCTION

A comprehensive preventive health care program should be a part of every older person's medical care. Numerous age-associated physiological changes and age-prevalent illnesses can be managed within a preventive health program that includes primary, secondary, and tertiary interventions. The goal of **primary intervention** is to prevent illness and maintain optimal function. The goal of **secondary intervention** is to detect disease at early stages. The goal of **tertiary intervention** is to prevent the progression and complications of disease.

II. PRIMARY INTERVENTION

A. Stop Smoking

Smoking increases the risk of lung cancer, cardiovascular disease, peripheral vascular disease, chronic obstructive pulmonary disease, and lower respiratory tract infections. Smoking also accelerates "normal" age-related physiological changes in pulmonary function.

B. Maintain Weight within 15% of Average for Age

Being overweight or underweight (15% over or under average
weight for age) may predispose to osteoarthritis, diabetes, hypertension, and difficulties in mobility. Maintenance of an adequate body
mass increases capacity to manage illness.

Caloric requirements decline with increasing age. This results
largely from a reduction in lean body mass. The basal energy
expenditure (BEE) appears to decrease in direct relationship to this
change.

The BEE can be calculated using the Harris–Benedict
equation:

Men: $(13.7 \times$ average weight in kg$) + (5.0 \times$ height in cm$) -$
$(6.8 \times$ age in years$) + 66$

Women: $(9.6 \times$ average weight in kg$) + (1.8 \times$ height in cm$) -$
$(4.7 \times$ age in years$) + 655$

The following factors have been suggested to better approximate total energy requirements:

Factor	Level of Physical Activity
1.1	Very light
1.2	Light
1.3	Moderate
1.5	Heavy

Chronic illness, however, may increase requirements to as high
as $2.5 \times$ BEE.

The following is a suggested range of caloric requirements for
weight maintenance as adjusted for age and sex:

Caloric Requirements (kcal/day)		
Age	Men	Women
51 to 75	2000 to 2800	1400 to 2200
76+	1650 to 2450	1200 to 2000

These requirements are increased by activity, surgery, and illness. Diets containing less than 1500 calories are usually vitamin and mineral deficient and require supplementation.

C. Insure Adequate Nutritional Intake

1. Protein

In general, a protein intake of 0.8 g/kg body weight is sufficient to maintain a positive nitrogen balance. Failure to do so increases the risk of infection, glucose intolerance, fatty infiltration of the liver, and altered neurotransmitter function.

2. Carbohydrates

Intake of refined carbohydrates should be limited. These foods are low in vitamin and fiber content and increase risk of dental caries. Carbohydrate should comprise 60 to 70% of most diets.

3. Fats

The American Heart Association recommends that people limit cholesterol intake to less than 300 mg per day. High-cholesterol-containing foods include eggs (250 mg/yolk) and organ meats including liver, kidney, and brain. Saturated to unsaturated fat ratio should approximate 1 : 3, and total fat intake should be limited to no more than 25 to 30% of calories.

4. Vitamins and Minerals

A well-balanced diet consisting of at least 1500 calories should provide under most circumstances adequate intake of vitamins and minerals. Although sodium intake should be minimized, recent data suggest that only a minority of elderly are "salt sensitive." Twenty-five percent of elderly people, when consuming a diet high in salt, increased their blood pressure. Pickles, TV dinners, canned foods, and cold cuts are foods often high in salt. In general, sodium intake should not exceed 5 to 7 g per day. Low-salt diets are expensive, hard to prepare, and often are not necessary. Although an adequate potassium intake is essential, caution is advised in those elderly with reduced renal function. Elderly persons rarely require supplemental iron; 1200 to 1500 mg of elemental calcium is the daily recommended requirement. Since many elderly do not consume a diet high enough in calcium, oral supplementation may be necessary. A well-balanced diet should contain the required 15 mg of zinc. Zinc has been shown to be essential to wound healing, prevention of decubiti, and immune function. Sources of dietary zinc include meat, chicken, and fish.

5. Fluid

Adequate hydration is essential to maintain normal renal and bowel function. Since a major fraction of fluid is derived from food sources, and caloric intake generally declines with age, oral fluid intake should be encouraged.

6. Fiber

Adequate fiber helps guard against constipation, hemorrhoids, and diverticulosis. Fiber must be introduced into the diet slowly to avoid intestinal symptoms and poor compliance. Adequate hydration is necessary to prevent fiber-induced constipation.

7. Alcohol

Alcohol should be taken only in moderation if at all. Changes in body composition that occur with age result decreased extra- and intracellular fluid compartments. This leads to a decreased potential volume of distribution for alcohol. The same intake of alcohol in the elderly as during youth may now result in a higher effective alcohol level, increasing the risk of falls, depression, and mental disturbances.

The onset of new alcoholism is a major problem in the elderly, particularly in men who have recently lost their spouses.

D. Maintain Psychosocial Needs

A comprehensive preventive health care program must insure that all psychosocial needs are met.

   1. Financial security is necessary to insure adequate nutrition, shelter, and medical care. Food stamps, medicaid, and relief organizations may need to be utilized.

2. A social network including family, friends, and colleagues is essential to optimal functioning.
3. Early signs and symptoms of psychological problems must be attended to immediately. Depression and dementia can be easily confused. Social isolation and loneliness must be prevented.
4. A safe, barrier-free environment should be established. A home inspection may be required. Just a few of the many environmental hazards that can present problems include loose rugs, inadequate lighting, stairs, and inappropriately placed electric cords and appliances.

E. Maintain Adequate Physical Activity

Physical conditioning is capable of retarding loss of bone mass, improving cardiopulmonary function, improving mobility, and benefiting the psychological profile. The elderly individual interested in embarking on an exercise program should be advised to have a complete physical examination including a resting electrocardiogram. Most agree that a stress test should be done if there is any question of cardiopulmonary reserve or if the expected program is to be a strenuous one. Exercise programs should include isotonic exercises consisting of full muscle movement without resistance. This is in contrast to isometric exercises wherein muscles are exerted against force. This may result in excessive increases in peripheral vascular resistance. Exercise should begin gradually and slowly be advanced in intensity. A heart rate of 60–80% of maximum as adjusted for age provides optimal cardiovascular conditioning. Studies suggest maximal effect when this exercise is performed three times per week. Although excessive exercise increases the risk of musculoskeletal problems, moderate exercise can be important in maintaining psychological well-being and in weight maintenance.

Suggested Guidelines for an Exercise Program in the Elderly*

1. In the beginning, the elderly patient may be discouraged by an increased amount of muscle and joint soreness. This is to be expected, and the patient should probably start with a more gradual regimen

---

(continued)

Suggested Guidelines for an Exercise Program in the Elderly* (*Continued*)

and understand that a conditioned effect takes time, extended effort, patience, and commitment.

2. Patients should avoid eating substantial amounts of food for approximately 2 hr before and 1 hr after exercise.
3. Elderly patients often have less ability to adapt to temperature extremes. Elderly people exercising in warm climates should be careful to guard against dehydration. Elderly exercising in cold climates should protect against hypothermia by wearing proper clothing (i.e., multiple layers of absorbent, preferably cotton, materials). The intensity of exercise should be decreased with climatic extremes.
4. All exercise sessions should be preceded with a gradual warm-up period and terminated with a full cool-down period.
5. Care should be taken not to take excessively hot showers or baths, since this can sometimes be deleterious to cardiac function and may precipitate a syncopal episode.
6. When ill, patients should refrain from exercising. Exercise may potentiate dehydration, possibly already present because of elevated body temperature and increased sweating.
7. Patients should know when to stop exercising: unusual discomfort, shortness of breath, chest pain, or palpitations require immediate exercise termination and medical consultation.
8. Often the elderly have impaired vision. Older adults who exercise should do so in well-lighted, flat-surfaced areas where the risk of falling is reduced. Elderly with impaired hearing should exercise in nontrafficked areas.

*Adapted from Adelman R. D., Greene M. G., Stewart M. M.: Preventive medicine for the elderly, in Gambert S. (ed): Contemporary Geriatric Medicine (vol 1), New York, Plenum Press, 1983, pp 371–396.

F. Periodic Health Evaluation

A yearly comprehensive medical evaluation is recommended for all persons over the age of 75; an examination every 2 years has been suggested for those aged 65–75. The evaluation should include a thorough history and physical examination including rectal, pelvic, and breast examination. More controversial is the need for yearly electrocardiogram and laboratory testing, including blood chemistry profile, thyroid function tests, and complete blood count. Thrice-yearly stool guaiac tests help detect occult colonic lesions. Yearly mammography has been recommended for all elderly women. An evaluation of the oral environment is required at least once a year, with many recommending every 6 months. Daily oral hygiene is essential.

G. Routine Immunization

1. Tetanus

Immunization against tetanus is necessary in the elderly as demonstrated by the 159 cases of tetanus found in people over age 50 in the United States during 1982 to 1984. A booster of tetanus toxoid (0.5 ml i.m.) is required every 10 years for all patients. Tetanus toxoid can lead to local pain and swelling as well as a severe reaction if given too often. Vaccination is contraindicated if a previous dose has led to a neurological reaction or hypersensitivity.

2. Influenza

Influenza vaccine can markedly reduce the incidence of complications, hospitalization, and death from influenza. The influenza vaccine, composed of inactivated whole virus or virus subunits grown in chick embryo cells, should be given annually (0.5 ml i.m.) to all patients with diabetes or other metabolic diseases; chronic pulmonary, cardiovascular, or renal disease; severe anemia; immunosuppressed patients; those in chronic care facilities; and all persons more than 65 years of age. Vaccination can rarely lead to fever, chills, malaise, and myalgia and is contraindicated in persons with an allergy to eggs.

3. Pneumococci

In persons with normal immune system function, vaccination against pneumococcus is approximately 60 to 75% effective in preventing the mortality associated with pneumococcal bacteremia.

The pneumococcal vaccine contains purified capsular polysaccharide from 23 types of *Streptococcus pneumoniae* that are responsible for 90% of recent bacteremic pneumoccal infections in this country. Vaccination should be given once during the lifespan (0.5 ml s.c. or i.m.) to all patients with alcoholism, cirrhosis, diabetes, nephrotic syndrome, renal failure, cerebrospinal fluid leaks, immunosuppression, Hodgkin's disease, chronic cardiac or pulmonary disease, and conditions that predispose to pneumoccal infection, particularly asplenism and sickle cell anemia, and to all persons more than 65 years of age. Vaccination can lead to local soreness and should not be given to persons who have received a vaccine within several years even if that vaccine contained fewer pneumococcal types. In addition, vaccination should not be given for at least 1 year following a pneumococcal infection. The pneumococcal vaccine can be administered at the same time as the influenza vaccine if different sites are used.

III. SECONDARY INTERVENTION

Medical evaluation should include a review of the following historical factors:

Topic	Comments
1. Medications	Review both prescription and over-the-counter medications for appropriateness of medications, dosage, side effects, drug–drug interactions, etc.
2. Immunizations	Obtain immunization records to insure adequate protection against tetanus, influenza, and pneumococcal infection.
3. Health habits	Inquire regarding exercise, nutrition, and other personal habits. Also review for evidence of malnutrition.

(*continued*)

Topic	Comments
4. Social functioning	Inquire into the adequacy of social support structure, i.e., family, friends, finances, housing.
5. Psychological functioning	Observe for signs or symptoms of a psychological disturbance, i.e., functional disturbance, dementia, sexual dysfunction, depression, or sleep disturbance.
6. Physical functioning	Determine ability to perform "instrumental" activities of daily living, i.e., writing, reading, cooking, cleaning, shopping, using telephone, managing money, and doing laundry. Determine ability to perform all or part of the six activities of daily living: bathing, dressing, toileting, transfer, continence, and feeding. Determine environmental barriers that may limit full function within the home or neighborhood.

The **physical examination** should include the following:

System	Evaluation
1. General	Evaluate height, weight, and skinfold thickness.
2. Integument	Assess for skin cancers; an examination of the feet is essential.
3. HEENT	Evaluate vision, hearing, and oral pharynx status.

(continued)

System	Evaluation
4. Cardiovascular	Monitor blood pressure; evaluate for orthostatic hypotension; evaluate for peripheral vascular disease.
5. Breasts	Examine breasts of both men and women.
6. GI	Guaiac stool test every year; periodic sigmoidoscopic or proctoscopic examination is suggested.
7. GU	The prostate should be examined for size and for masses.
8. Gynecological	A pelvic examination is suggested during each health evaluation; the Pap smear is thought to have a low yield in women over 65 years of age.
9. Musculoskeletal	Observe for signs of contractures, decubiti, and muscular weakness.
10. Neurological	Assess for neuropathy, gait disturbance, and coordination.
11. Endocrine	Palpate thyroid gland and evaluate for signs of hyper- or hypothyroidism.

Although the need for regular **laboratory evaluation** remains controversial, one or more of the following tests have been recommended at periodic intervals:

1. Hemoglobin and hematocrit.
2. Mean corpuscular volume or evaluation of peripheral smear.
3. White blood count.
4. Electrolytes.
5. Glucose.
6. Blood urea nitrogen and creatinine.

7. Liver function tests.
8. Cholesterol and albumin.
9. Thyroid function tests.
10. Electrocardiogram.
11. Chest roentgenogram.
12. Tuberculin or other TB skin test.
13. Urinalysis.
14. Mammogram.

IV. TERTIARY INTERVENTION

A tertiary prevention program is best accomplished by working closely with the elderly person or care giver to develop a plan that insures optimal physiological, psychological, and social functioning. The elderly person or care giver should be able to anticipate the natural progression of the condition and be able to identify when professional help is necessary.

Examples of tertiary prevention programs for specific problems follow.

A. Physical Disability

1. Prescribe an individualized restorative care program that strives to optimize function.
2. Following completion of the restorative program, therapy should be initiated to maintain optimal function.
3. Barriers within the home should be eliminated and adaptive equipment introduced, e.g., railings, ramps.
4. Arrangements should be made to seek alternative living quarters if the home is inaccessible.
5. Arrange special services to assist the physically disabled, e.g., meals-on-wheels, home health aid, visiting nurse.
6. Facilitate social contact to avoid isolation and loneliness, e.g., transportation services, visitors, contact with family and friends.

B. Physical Instability

1. Review medications for their potential to cause falls.
2. Rule out illnesses predisposing to falls, e.g., orthostatic

hypotension, syncope, vertigo, Parkinson's disease, seizures, and cerebrovascular disease.
3. Treat visual disturbances that predispose to falls, e.g., cataracts, presbyopia, glaucoma, and macular degeneration.
4. Counsel against substance abuse, e.g., alcohol, barbiturates.
5. Periodically evaluate for environmental hazards, e.g., inadequate lighting, waxed floors, torn or frayed rugs, broken stairs, electric cords, icy sidewalks, high-crime neighborhoods.
6. Install safety measures, i.e., railings, nonsliding rugs, alarms, etc.
7. Prescribe a four-point cane or walker if an unsteady gait is noted.

C. Malnutrition

1. Evaluate whether any age-related change in physiology is contributing to the malnutrition, e.g., diminished senses, altered gastrointestional function.
2. Illness predisposing to malnutrition must be promptly treated, e.g., dental caries, depression, infections, cancers.
3. Rule out malnutrition related to medication use, e.g., anorexia resulting from digitalis intoxication, hypokalemia resulting from diuretics, hyperkalemia resulting from use of potassium-sparing diuretics, vitamin B deficiency following use of isoniazid and L-dopa, magnesium deficiency resulting from diuretic use, deficiency of calcium, phosphorus, and/or fluoride resulting from use of aluminum-containing antacids, electrolyte disturbance, weight loss, and vitamin A, D, E, and K deficiency resulting from laxative abuse, and vitamin D and folate deficiency resulting from phenytoin use.
4. Employ a 24 to 48-hr recall or 48 to 72-hr food diary to determine actual food intake. The 24-hr recall, however, is often not accurate, especially in the cognitively impaired person. Despite this limitation in quantification, types of food given can help assess quality of diet.
5. Suggest ways to modify the diet in order to meet the recommended requirements.
6. Continue to monitor for nutritional adequacy using an-

thropometric measurements, physical examination, and
laboratory testing.

7. Encourage group feeding programs (senior citizen groups,
church groups, etc.) if social isolation is interfering with
caloric intake.
8. Suggest alternative ways to improve food palatability.
9. Suggest use of food stamps, application for medicaid, or
help from community group if finances are too limited to
purchase or prepare an adequate diet.
10. Arrange assistance if needed for shopping and/or cooking.
11. Suggest alternative meal arrangements if illness or limited
attention span interferes with caloric intake, i.e., smaller
and more frequent meals, meal supplements, etc.

D. Dementia

1. Perform a thorough history, physical examination, and
laboratory evaluation to rule out reversible causes of
dementia.
2. Stress the need for family members to consult their
physician if there is an acute change in mental status or
other signs and symptoms of acute illness.
3. Continue health maintainance and health promotion as
outlined in Sections II and III.
4. Inform the family or care giver of what they might expect
during the clinical course of the disease, including the not
uncommon problems of wandering, incontinence, sleep
disturbance, and physical and verbal abusiveness.
5. Suggest a safe, structured environment; reassurance and
repeated reality orientation should be encouraged, i.e.,
with signs, familiar surroundings and routines, and verbal
cues.
6. Encourage ways to combat altered sleeping patterns. Sleep
can be facilitated by keeping the demented individual up
and more active during the day and by minimizing daily
naps and visits to their rooms.
7. Encourage the spouse and or family members to join a
support group and/or obtain information relating to their
specific problem.
8. Prescribe psychotropic medication only if problems such
as agitation or wandering cannot be managed through

nonpharmacological means. The psychotropic medication should be carefully titrated. Medications with low anticholinergic properties are better tolerated by the elderly and have fewer side effects including sedation, areflexic bladder, bowel atony, cardiac arrhythmia, and reduced salivary flow.

9. Be watchful for family stress. "Respite" programs may be necessary to maintain the demented person at home long term. When necessary, advise regarding nursing home placement. Support families during these stressful times.

10. Observe for signs of elder abuse.

SUGGESTED READING

American College of Physicians: *Guide for Adult Immunization*. Philadelphia, American College of Physicians, 1985.

Canadian Task Force on the Periodic Health Examination: Periodic health examination. *Can Med Assoc J* 121:1193–1254, 1979.

Centers for Disease Control: Adult immunization, recommendations of the immunization practices advisory committee. *Morbid Mortal Week Rep* 33:1S, 1984.

Centers for Disease Control: Tetanus–United States, 1982–1984, *Morbid Mortal Week Rep* 34:602–611, 1985.

Kane R. L., Kane R. A., Arnold S. B.: Prevention and the elderly: Risk factors. *Health Serv Res* 19:945–1006, 1985.

15

Patient Education Handouts

HELPFUL HINTS TO AVOID CONSTIPATION

1. There is no "normal" number of times one should expect to have a bowel movement each day. **Do not** be preoccupied with your bowel habits.
2. Report any change in bowel habits to your physician.
3. Stay active! Regular exercise increases bowel motility.
4. Drink plenty of liquids, 1 to 2 quarts per day, unless you have heart, circulatory, or kidney problems. In this case, discuss intake with your physician.
5. Do not neglect the "urge" to defecate.
6. Comfortable and familiar toilet facilities may improve bowel function.
7. Try emptying your bowels at a preset time each day.
8. Take advantage of the normal "gastrocolic reflex"; try emptying your bowels 10 to 20 min after a hot drink, breakfast, or dinner.
9. Allow adequate time for bowel movements.
10. Foods containing fiber (15 to 30 g per day) improve bowel function.
11. Try eating fewer highly processed foods such as sweets and fewer foods that are high in fat.
12. Avoid use of laxatives and/or enemas.
13. If bowel movements continue to be a problem, see your physician.

From *Handbook of Geriatrics*, Steven R. Gambert (editor), Plenum Press, New York, © 1987. May be reproduced for distribution to patients.

HELPFUL HINTS THAT MAY IMPROVE MEMORY

1. **Be alert and aware.** Anything you wish to remember you must first observe carefully. When you really pay attention, you will become aware of things that ordinarily might make only a vague impression. Since concentration is essential to improving memory, make sure that you are concentrating on one thing at a time and that all forms of distraction are minimized.

2. **Link ideas to images.** All memory is based on associating new information to something that you already know. In improving memory, the trick is to link what you want to remember to a strong visual image. For example, most people easily remember the shape of Italy because they have been told that Italy is shaped like a boot. Similarly, association can help with names. Say you want to remember the name Chandler. By visualizing a specific chandelier you already know, you will be more likely to remember the name.

3. **Cure absentmindedness.** Forgetting to turn an oven off or misplacing a set of keys is basically a memory problem. The cure is the trick of association. A timer can help remind you when to turn the oven off, and associating important items, such as keys, with specific places in the home can be helpful. For example, you might hang keys on the doorknob. Thus, you always associate the doorknob with your keys.

4. **Be orderly.** Forming good habits around the home can help to counteract memory lapses. Medications or dentures should always be kept in the same place. Making lists of things to remember is helpful, as are calendars for remembering birthdays and anniversaries.

5. **Repeat three times.** Studies have shown that for most people a new piece of information must be repeated at least three times before it becomes fixed in the memory bank. Some people need as many as 16 repetitions, irrespective of their age. For an even stronger impression of the memory, the idea should be repeated verbally and linked to some visual image.

From *Handbook of Geriatrics*, Steven R. Gambert (editor), Plenum Press, New York, © 1987. May be reproduced for distribution to patients.

6. **Organize new information.** New information can be remembered more easily if it is organized into categories. Either a mental or written outline can serve as a memory cue. A speech or a lesson can always be remembered more easily if it is arranged in a logical order with particular subheadings. Organization can also involve words that help you remember. For example, to remember the names of the five Great Lakes, some people picture **HOMES** on the lake: Huron, Ontario, Michigan, Erie, and Superior.

7. **Relax and take your time.** Memory functions best when you are relaxed and can concentrate. When you are tense, tired, or emotionally upset, you cannot expect optimal memory function. Older people cannot absorb too much new information at one time, and they may take more time to do so than a younger person. For them, learning just a little at a time may be helpful. If you are unable to recall relatively unimportant information, you should try to avoid anxiety. If you take your mind off it, it very likely will come to you at a later time.

From *Handbook of Geriatrics*, Steven R. Gambert (editor), Plenum Press, New York, © 1987. May be reproduced for distribution to patients.

MAKING YOUR KITCHEN A SAFER PLACE TO BE IN

1. Place all shelves at eye level or put commonly used foods on the counter. Items in the refrigerator should be stored within easy access.

2. Avoid excessive bending or reaching. Use of chairs and ladders can be dangerous. Stepstools must be sturdy and used according to specification.

3. All chairs should have high backs, armrests, and non-slip legs. The seat should be of appropriate height to allow easy ability to get on and off the chair.

4. Provide water-absorbing mats around sinks.

5. Use nonskid floor wax.

6. Clean up spills promptly.

7. Use shoes with nonskid soles.

8. Install smoke detectors in the kitchen.

9. Electric stoves are preferred over gas stoves.

10. A fire extinguisher should be within easy reach.

11. A supply of baking soda should be within easy reach to help put out fires.

12. Have emergency telephone numbers pasted near the kitchen telephone.

13. Have adequate lighting in the kitchen, especially around cooking and cutting areas.

14. Do not wear long or loose clothing in the kitchen. Clothing should be made of noninflammable materials. Plastic aprons can be hazardous.

15. Clearly mark "on" and "off" positions on all kitchen appliances. Red paint or fingernail polish can be used.

16. Store all appliances, utensils, and cooking accessories in secured and marked areas.

17. Pot holder mittens should be used when picking up hot items.

From *Handbook of Geriatrics*, Steven R. Gambert (editor), Plenum Press, New York, © 1987. May be reproduced for distribution to patients.

APPROXIMATE FIBER CONTENT OF SELECTED FOODS*

Grains and cereals		Fruits	
White bread	2.7	Prunes	7.7
Whole wheat		Banana	3.4
bread	8.5	Raisins	6.8
Pancake, waffle	0.9	Apple (peel and	
All-Bran™	26.7	flesh)	1.5
Cornflakes	11.0	Cherries	1.2
Rice Krispies™	4.5	Dried apricots	24.0
Special K™	5.5	Orange	2.0
Puffed wheat	15.4		
White rice	0.8		
Oatmeal	7.0		
Vegetables		Others	
Peas	12.0	Peanuts	8.1
Spinach	6.3	Peanut butter	7.6
Green beans	3.2	French fries	3.2
Corn on cob	4.7	Lentil soup	2.2
Cauliflower	1.8	Strawberry jam	1.1
Broccoli	4.1	Bulk laxatives	
Baked potato	2.5	Metamucil® (g/tsp)	3.5
Baked beans	7.3	Perdiem Plain®	
Lettuce	1.5	(g/tsp)	5.2
Cucumber	0.4	Fiber Med®	
Onions	2.1	(g/cookie)	5.0
Carrots	2.9		
Celery	1.8		

*grams fiber/100 g edible portion.

From *Handbook of Geriatrics*, Steven R. Gambert (editor), Plenum Press, New York, © 1987. May be reproduced for distribution to patients.

CALCIUM EQUIVALENTS

Quantities of foods needed to supply the amount of elemental calcium (291 mg) in one cup (8 oz) of whole milk.

Food Source	Approximate Measure	
Roast beef	5	lb
Eggs	10	
Peanut butter	29	tbsp
Salmon with bones and oil	4	oz
Sardines	2¼	oz
Tuna	7¾	lb
Rice	14	cups
Oatmeal	13	cups
Bread (white) slices	13–15	
Cornflakes	73	cups
Egg noodles	18	cups
Apples	29	
Bananas	29	
Corn	36	cups
Potatoes, baked	22	
Greens (collards, kale, turnip, mustard)	1	cup
Broccoli	2	cups
Yogurt	1	cup
Pudding	1	cup
Milk, sweetened condensed	⅓	cup
Milk, evaporated	½	cup
Ice cream	1¾	cup
Buttermilk	1	cup
Cheese (American, cheddar)	1½	oz
Cottage cheese (creamed and low fat)	2	cups
Cottage cheese (dry curd)	6	cups
Swiss cheese	1	oz
Cheese spread	2	oz

From *Handbook of Geriatrics*, Steven R. Gambert (editor), Plenum Press, New York, © 1987. May be reproduced for distribution to patients.

AVOIDING ACCIDENTS AT HOME

1. Illuminate all stairways and provide light switches at both top and bottom.
2. Provide nightlights or bedside light switches.
3. Stair handrails should be secure and at proper height.
4. Carpets must be tacked down.
5. Nonskid treads are recommended for stairs.
6. The use of "throw rugs" is discouraged.
7. Furniture and other objects must be arranged so that they do not obstruct frequently traveled pathways.
8. Nonskid mats or strips should be used in the bathtub.
9. Handrails may be useful near toilet fixtures.
10. Outdoor steps and walkways must be well lighted and in good repair.
11. Never smoke in bed or when tired.
12. When in the kitchen, do not wear flammable, long, and/or loose clothing.
13. Avoid having water heater thermostats set at "scalding" levels.
14. Fire/smoke detectors should be installed at appropriate places.
15. An emergency exit route should be planned and well understood by all household members.
16. Locks should be secure yet easy to open in times of emergency.
17. Keep emergency telephone numbers readily accessible.
18. Do not climb on ladders, tables, or chairs.
19. Use stepstools only according to specification and only if not alone.
20. Use only nonskid flooring and keep off wet floors.
21. Shoes must be well secured and low heeled. Long clothing may also result in falls.

From *Handbook of Geriatrics*, Steven R. Gambert (editor), Plenum Press, New York, © 1987. May be reproduced for distribution to patients.

HOW TO PREPARE YOURSELF AGAINST COLD WEATHER

1. Air should be humidified. This not only makes you feel warmer but also helps prevent dry mucous membranes.
2. See your physician regarding vaccinations against influenza and pneumococcal pneumonia.
3. Keep room temperature between 70° and 72°F.
4. Keep an adequate supply of nonperishable food in the house.
5. Wear "layers" of clothing.
6. Wear a hat; the head is a major source of heat loss.
7. Natural fibers are good absorbants of perspiration.
8. Do not shovel snow or use a snow blower.
9. Wear nonslip, flat shoes when outside.
10. Avoid walking on snow or ice.
11. Any change in appetite, sleeping pattern, or general health must be promptly reported to your physician.
12. All heating equipment must be carefully inspected and used only according to specifications.
13. Thermostats should be clearly labeled. Red paint or nail polish can be used to mark preset temperatures.
14. Elderly persons who live alone should be contacted regularly.
15. Visits by family, friends, and/or neighbors should be encouraged.

From *Handbook of Geriatrics*, Steven R. Gambert (editor), Plenum Press, New York, © 1987. May be reproduced for distribution to patients.

Index